Puzzles to Unravel the Universe

By Cumrun Vafa

To my beloved wife and lifetime friend,

Afarin,

and to my dear sons,

Farzan, Keyon and Neekon,

who were my inspiration for writing this book

and to my loving parents, Simeen and Javad,

for nurturing my curiosity.

"Puzzles to Unravel the Universe *is a wonderful survey of concepts central to modern physics and mathematics explored through the medium of puzzles. This is one of the most unusual and captivating approaches I have ever encountered, allowing the reader–whether novice or expert–to learn through the enjoyment of solving brain-teasers. What a fun and effective way to actively absorb essential and cutting-edge ideas from one of the world's greatest physicists."* –Brian Greene

"This book offers a fascinating and unusual tour of advanced ideas in physics and mathematics, illustrated with elementary and entertaining puzzles. Readers will find much to enjoy and learn from." –Edward Witten

Contents

Preface

We have an innate interest in understanding how things work. We hope to observe patterns around us that help us anticipate what comes next. Quantifying these patterns is what gradually led humans to develop mathematics. It is therefore not surprising that mathematics is the natural language for describing how nature works. Indeed, mathematics is the backbone of physics, which aims to describe how the universe works at its most fundamental level. The deeper we understand the laws of nature, the more we need advanced topics in mathematics, so much so that today physics has a reputation of being impenetrable by the uninitiated due to its mathematical complexity.

However, this perception overlooks the simplicity of physical laws and the elegance of mathematics in capturing the main essence of physical reality. As a physicist with a keen interest in mathematics, I have witnessed firsthand how beneath all the complex and formidable looking mathematical structures appearing in formulating physical laws, there lie simple and deep nuggets of truth. These truths are what many scientists strive to crystallize when the dust settles and the laws of nature have been discovered. These nuggets are a kind of "executive summary" that scientists hold dear as takeaway lessons from the discoveries about the laws of nature. Fortunately, these core ideas can often be illustrated with simple mathematical puzzles. These puzzles are so simplified that one does not need any extensive background in physics or mathematics to tackle them and appreciate their meaning. Mathematical puzzles of this type are not only great fun to work on, but deeply satisfying because they capture some deeper meaning about physical reality, beyond just a puzzle. It is my aim in this book to take the reader on a journey to unravel aspects of the laws of the universe through fun puzzles.

The common theme in this book is the idea that beneath physical reality there exists not a single overarching idea, but rather a collection of almost opposing concepts which together frame physical reality. An appreciation of how these opposing thoughts can weave together and work in harmony towards a remarkable end is the main aim of this book. I hope to demonstrate these concepts by viewing some of the most important principles discovered about nature through the prism of puzzles.

After briefly reviewing the history of science and the interaction between mathematics and physics over centuries, I turn to the main topics one-by-one. Each section of the book starts with a thought in one subject and then discusses the importance of the opposite thought. And then the same is repeated by switching subjects between physics and mathematics. All of it is presented through the backdrop of fun puzzles.

The first topic is symmetry. On the one hand, we see the significance of preserving symmetry in mathematics and physics; on the other, we cover the importance of breaking symmetries. The puzzle of designing the shortest highway connecting four cities on the corners of a square is a beautiful example of this phenomenon. While symmetries explain how conservation laws like the conservation of energy works, we will see why breaking symmetries is even more important to our very existence. As we discuss, this is related to the recently discovered Higgs particle. I also explain how our own eyes and their location on our face manifests a breaking of symmetries. We discuss the importance of both intuitive and unintuitive ideas in both physics and mathematics. Intuitive ideas (such as continuity featuring prominently in various aspects of physical laws) and unintuitive abstractions (such as viewing time as an extra dimension) are necessary to more deeply understand reality. We show that the idea of continuity, as simple as it is, leads to powerful conclusions. An illustration is the puzzle that reveals why there are always diametrically opposite points on the equator with the same temperature. We also show how continuity of the laws of physics can explain why Albert Einstein's theory of general relativity predicts that there are always an odd number of gravitational images of a star. We then turn to the idea of naturalness: how to make ballpark estimates for how nature works with very little information. For example, we show through a simple estimate how much we need to shrink the sun in order to make it a black hole. We then turn to the opposite thought and discuss how unnaturally large or small numbers appear in the fundamental laws of nature, which are hard to anticipate. In particular, why is the gravitational force between protons a trillion, trillion, trillion times smaller than the electric repulsion between them? We illustrate the appearance of unexpectedly large numbers in physics by Archimedes' old cattle problem, whose solution involves a number close to a million digits! I also venture to briefly discuss some of the connections between science and religion, but unlike what one often sees in the context of this discussion, even this topic is framed in the language of fun puzzles. One such example is a puzzle which involves a rectangle comprised of smaller ones. Each side of the rectangle has integer length, leading to the same property for the bigger rectangle. Finally I discuss some of the most exciting modern developments in fundamental physics, in the context of string theory. String theory has recently emerged as a unified quantum theory encompassing all the fundamental forces. I focus on the idea of duality in string theory, which has fascinated string theorists for the past couple of decades and has played a key role in its development. I discuss how duality, for example, leads to a better understanding of black holes and the nature of space and time. A puzzle that illustrates duality is the colliding ants on the meter stick, where each ant aims to keep from falling off the edges for as long as possible. It turns out that the idea of duality discovered in string theory is a microcosm of this book: It is the idea of how opposite principles can fit seamlessly in a consistent and powerful way to predict how nature works. Nothing can be more powerful than having opposite thoughts work in harmony, which is why duality has become a most powerful tool in unraveling the deepest secrets of our universe.

I hope you find reading this book and working through its puzzles to be interesting and edifying. I will be pleased if you find a new appreciation for the fundamental laws of our universe and how mathematics fits in it, while gaining an appreciation for the power of puzzles to challenge and inform us, and sometimes even surprise us! And even if you were not a puzzle lover as a kid–as I was and still am–it is never too late to become one!

I have been fortunate enough to take some of the freshmen of Harvard College on this journey of discovering how puzzles can unravel mysteries of the universe through a seminar I designed to this end. This book is a result of this course and has been enriched by the feedback and suggestions I received from students who took it. This book was based, initially, on notes taken by three students–Tony Feng, Kewei Li, and Weiming Zhao, and were substantially edited by Steve Nadis. Some figures were added by Xiaotian Yin. I have also benefited from encouragement of a number of colleagues and in particular Yaotian Fu and Brian Greene in completing this book. To all of them, I am deeply grateful. I am sure there are many ways this book can be improved. If the readers have any suggestions, I would be happy to receive them through my webpage, www.cumrunvafa.org .

And last, but not least, it was my wife Afarin's suggestion that led me to develop this course and write this book. This book would not exist without her enthusiasm for this project. I am deeply grateful to her.

1. An Introduction to Modern Physics

Many fundamental aspects of physics have simple mathematical underpinnings that may be concealed by the complexity of the formalism–both the unfamiliar language and the sometimes daunting equations. The same is true of many abstract mathematical ideas, which often involve simple concepts that can, nevertheless, get obscured due to the setting in which they are presented. Deep ideas in physics and mathematics often share a common core, which is perhaps not surprising given the closeness of these two disciplines. More surprising is the fact that some of those same ideas can emerge in the course of solving mathematical puzzles.

This book is about puzzles and their relations to mathematics and physics. While puzzles can be fascinating and entertaining in and of themselves, we are going to see how they can serve as a bridge between the two fields and reveal some of the common bonds they share. It does not take an advanced knowledge of mathematics and physics to solve the puzzles presented in this book, nor is it assumed that the reader has a deep background in either field. But a keen interest, as well as some training, in those subjects would certainly be helpful in appreciating this book, with college students or advanced high school students being among the intended audience.

Although physics and mathematics are tightly intertwined, they have very different cultures and philosophies. Mathematics starts from fundamental axioms, drawing upon logical inferences to build from there. Physical laws, rather than being logically deduced in a hierarchical fashion, have been developed to explain how different parts of nature work and how the laws of nature fit together. Physics stresses the *relationships* between these laws, rather than their logical dependencies. Of course, logical cohesion of these ideas is still a necessary ingredient of physical laws. In mathematics, it is important to be clear as to what your underlying axioms and assumptions are. On the other hand, as we will see later, the axioms, or fundamental principles, of physics can change as new evidence or theoretical ideas come to light.

History shows that important progress in the field can occur when what was first regarded as a consequence of a physical law is subsequently determined to be a guiding principle in its own right. A good physicist, therefore, should always be open to revisions or "reshufflings" of this sort because the newly recognized principle often ends up being more fundamental, and having a wider range of applicability, than the original principle from which it was initially thought to have stemmed. The principle of conservation of momentum offers a good example. Although it was first viewed as a consequence of Newton's laws of motion, it was subsequently determined–more than 225 years after Newton's laws were presented in *Principia Mathematica*–that conservation laws were more fundamental than laws of motion because they stem from underlying symmetries of nature.

It is for this reason that physicists try to maintain a flexible attitude as to what the fundamental principles are, which is constantly evolving. Rather than attaching too much value to the hierarchical nature of ideas, physicists are willing to reorder the structure at any moment, which is contrary to the way mathematicians usually view mathematics. A mathematical theorem, if proven correct, is regarded as eternally true–unlike physical principles that are subject to change as new empirical findings come to the fore.

There are other differences, as well. Explaining complicated phenomena in physics, for example, often involves the kinds of approximations that mathematicians may be loath to make. The question, for instance, of whether a space is "continuous," containing no gaps whatsoever, or is instead made up of points that are close together, may not matter to physicists focusing on the outcome of experiments carried out over much larger distance scales. For mathematicians, on the other hand, the smoothness of a given space or lack thereof, is a key feature, rather than being an irrelevant distraction.

The goal for this chapter is to provide a brief overview of the landscape of physics. This will be a quick review that is light on exposition; the aim here is not to be comprehensive, as that is essentially impossible in the course of a single chapter. Instead, we intend to touch on a few examples from the history of physics that can provide a sense of where we are today in the longstanding quest to understand the fundamental laws of nature.

Ancient Thoughts

The Greeks had many intriguing ideas about physics in their attempt to explain what was happening in the world around them. They were enamored with elegant mathematics, and some scholars, including Plato, believed that the truth of this world lies in geometry. They saw beauty in Euclidean geometry and Platonic solids, which they thought could be used to describe nature as a whole. While most of their work on mathematics was ages ahead of its time, their physics was not at the same level. Aristotle, for example, believed that rocks fall downwards because they like being on the ground. Among all possible states, he affirmed, being on the ground is the one that rocks enjoyed most. Consequently, he went on to explain, rocks fall faster as they approach the ground because they are happy to be nearer to their natural and preferred resting place[1].

[1]see *On the Heavens* by Aristotle

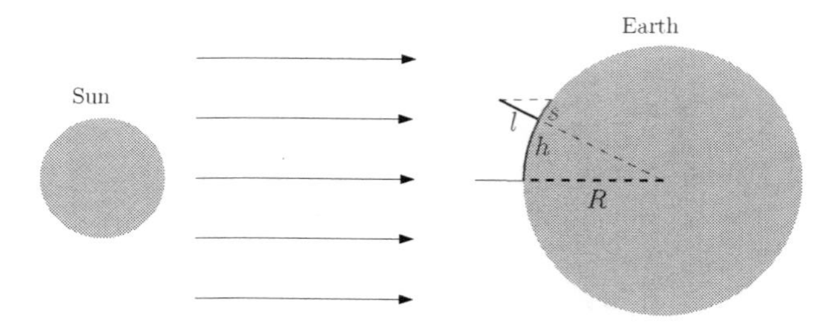

FIGURE 1. The circumference of Earth was measured by Eratosthenes of Cyrene around 230 B.C.

Despite the inadequate descriptions of physical phenomena by the ancient Greeks, their basic aspiration to describe the world with beautiful mathematics remains vital to science even today. Some of their thoughts, such as the notion of matter being comprised of individual atoms (advanced by Leucippus and Democritus, among others), remain accurate to this day. They not only believed Earth was a sphere, they also measured the planet's circumference sometime around 230 B.C. Eratosthenes, in particular, took simple ideas of trigonometry, along with observations of how the length of a shadow changes as we move a certain distance from the equator, in order to calculate the radius of the Earth. The answer he got was not too far off–about 15% of the actual radius of the Earth as measured today. The basic idea he employed is that as you go a distance h above the equator, the shadow of a stick of length l at noon grows from 0 to s (see Fig.1). The radius of the Earth, R, can then be deduced from simple trigonometry to be

$$R \sim h \cdot \frac{l}{s}$$

The idea of applying notions from pure geometry to deduce interesting facts about nature has continued long past the time of the early Greek mathematicians. Around 1000 A.D., Ibn Muadh and Ibn Al-Haytham determined that the height of the atmosphere was around 52 miles[2]), which is within 20% of the accepted value today. Muadh and some other Muslim scientists used the angle of depression of the sun at twilight and simple trigonometric functions to make his calculation. The approach was rather simple: the reason the sky does not get immediately dark after the sun sets, he argued, must be because the upper parts of the atmosphere can still receive light from the sun, even after sunset (see Fig.2). Measuring how long (t) it takes for sunlight to "run out," which is a couple of hours, as a fraction of the length of day, Muadh reasoned, is related to the height of the atmosphere h as a fraction of R the radius of Earth ($\frac{1}{2}(\frac{t}{24})^2 \sim \frac{h}{R}$).

[2]see `http://link.springer.com/article/10.1007%2FBF02464977`

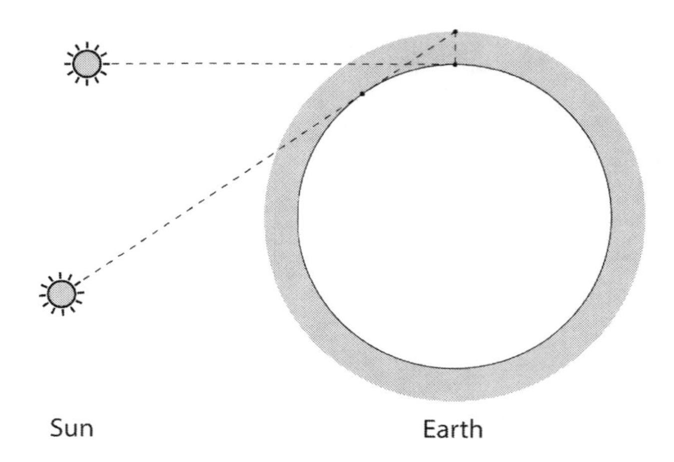

FIGURE 2. The height of the atmosphere was measured by Ibn Muadh and Ibn Al-Haytham in the 11th and 12th century.

But deep applications of mathematics to physics had to await more modern times, and the work of Sir Isaac Newton in the mid-to-late 1600s may constitute a true starting point in this regard.

Newtonian Mechanics

Newton was, without question, one of the great pioneers of modern physics. His second law of motion is encapsulated in one of the most famous equations of physics, which describes a differential relationship between the *position* $x(t)$ and the *force* F:

$$a := \frac{d^2x}{dt^2} = \frac{F}{m}.$$

While F and m are physical quantities, the *acceleration* a is more of a mathematical quantity, defined as the second derivative of position with respect to time. As physics became more quantitative, mathematics became increasingly enmeshed with physics. In fact, Newton had to invent a whole field of mathematics, calculus, in order to formalize his second law in precise mathematical terms. This is just one of numerous examples where the requirements for expressing the laws of physics prompted the development of new branches of mathematics. Mathematics, conversely, has also led to new insights in physics. Throughout the book, we are going to see much more of this back-and-forth interconnection and give-and-take between the two fields.

Lagrangian and Hamiltonian Mechanics

The continued exploration of the mathematical underpinnings of Newtonian mechanics, as examined in different physical contexts, led to its reformulation–along with some new mathematics, as well. In the late 1700s, for example, Joseph Louis-Lagrange suggested a new, so-called "Lagrangian" way of framing of mechanics, which produced the same physical results as Newtonian mechanics but was based on the "principle of least action" instead of force. Action is an integral that can be defined for every possible path a particle could take from its starting point to an ending point, given by the equation, $S = \int(K - V)\,dt$, where K is the kinetic energy of the particle and V is the potential energy of the particle along the path (see Fig.3).

The principle of least action states that the path the particle will actually take is the one that minimizes the action. If there are multiple solutions, each solution will yield an extremum, a minimum or maximum, of the action.

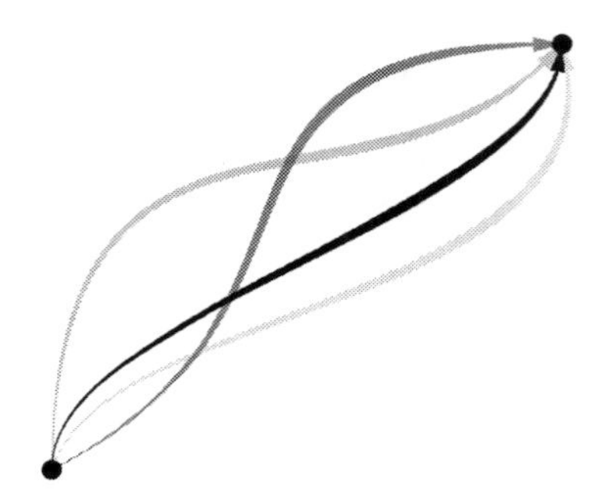

FIGURE 3. The Lagrangian formulation of mechanics looks at all possible paths from a starting point to an ending point. The physical path–or the path that would naturally be followed–is the one that minimizes a quantity called action.

This new way of looking at things made it easier for physicists to study mechanics under constraints, such as a ball rolling down a hill of a given topography or for a spinning top on different kinds of surfaces. To formalize Lagrangian mechanics, Euler and Lagrange invented an entire field of mathematics called the calculus of variations, which deals with extremizing integrals along paths, the solution of which satisfies the Euler-Lagrange equation. Note that this is more complicated than finding minima of a function of a finite number of variables, because there are infinitely many paths that connect two points in space. So in a sense it is equivalent to finding the minimum of a function (the action) of infinitely many variables (constituting the space of all paths). Physicists could use the calculus of variations, in turn, to pick out the path with the minimum possible length. The revamping of classical mechanics, made possible by the ideas of Lagrange and Euler, set the stage for future tie-ins with 20th-century physics–quantum mechanics, in particular, which was not readily accessible through the equations originally set down by Newton.

In yet another reformulation of classical mechanics, Hamilton reduced second order derivatives in time to first order derivatives by using twice the number of variables. Hamilton viewed both the position function and the momentum function–$x(t)$ and $p(t) = mv(t)$–as fundamental variables, rather than considering $x(t)$ on its own as had traditionally been done. Hamiltonian mechanics, as the new terminology was called, marked the beginning of the modern notion of doubled space, or phase space–the space defined by position and momentum. Hamiltonian mechanics turns out to be useful in quantum mechanics, as we will discuss below. Today we regard the Lagrangian and Hamiltonian formulations of mechanics as being more general and more fundamental than Newton's laws and thus have a wider range of applicability. This is an illustration of the fact that the axioms of physics are not immutable, nor is its underlying framework. Both can and do change over time.

Maxwell's Electromagnetism

When Maxwell began developing his theory of electromagnetism, many of its separate aspects had already been understood by Michael Faraday and others. In attempting to unite the different laws, Maxwell uncovered a *mathematical* inconsistency between the different equations, which he resolved by adding a new mathematical term (now called Maxwell's term) to his equations. This term was difficult to measure in the laboratory, but he did notice that it implied the existence of waves made up of electric and magnetic fields, moving at a speed that his equations predicted to be close to what was then estimated as the speed of light. This inspired Maxwell to postulate that light *is* nothing but an electromagnetic wave![3] This was yet another demonstration of the power of mathematical logic to predict new physical phenomena: Maxwell's correction arose from *mathematical* rather than physical considerations. His discovery of a simple mathematical inconsistency led him to conclude that light is made up of electric and magnetic disturbances moving through space–a triumph of human thought! This is one among countless examples showing that mathematical principles can be sufficient to motivate new physical laws.

Maxwell's equations in empty space lead to equations of the form

$$\frac{\partial^2 \vec{F}}{\partial t^2} = c^2 \nabla^2 \vec{F}$$

where $\vec{F}$ can be either the magnetic field $\vec{B}$ or the electric field $\vec{E}$. The solutions to this equation yield electromagnetic waves, which move with speed c, the speed of light.

That was not quite the end of the story, as further questions were raised. If you use this equation, you find that the wave will indeed travel with a speed c. But how, exactly, is that measured? Are we talking about the speed with respect to the

[3]It was further postulated that this wave required a medium, then referred to as the luminiferous aether. (See *A Guide to the Scientific Knowledge of Things Familiar* by Brewer for an indication of scientific thinking at the time), which was later disproved by the Michelson-Morley experiment.

Earth? Or the sun? And which kind of observer does this equation pertain to–just stationary ones, or does it hold for moving ones as well? In particular, if we move with constant velocity relative to an inertial frame, Newton's laws still apply, but the speed that such moving observers would measure for the electromagnetic waves would naturally be different. It was believed at the time that c could not be the same for all inertial frames because that would contradict the velocity addition law in Newtonian mechanics. In other words, Maxwell's equations lacked the symmetry of Newtonian mechanics (Galilean symmetries) which tells you that velocities change when you change the inertial frames depending on the relative velocity of the frames. So Maxwell's great insight seemed, at first glance, to lead to a contradiction.

Hendrik Lorentz then stepped in, offering a mathematical way to discover symmetries of Maxwell's equations which were different from those expected based on Newtonian mechanics. The Lorentz transformations teach us how the electric and magnetic fields as well as position (x, y, z) and time t change as we go from one frame to another. That, in turn, allowed the equations to look the same for all inertial frames. In other words, this led to the *Lorentz transformations*, as opposed to the Galilean transformations used in Newtonian mechanics. But this formulation had bizarre physical implications, such as the Lorentz contraction, the phenomenon that lengths *shrink* when shifting between inertial frames. Lorentz noticed, in particular, that in order to make Maxwell's equations work, regardless of the speed of an observer with respect to an inertial frame of reference, he had to make lengths shrink. Lorentz had difficulty making sense of this finding, trying unsuccessfully to explain the effect by means of electric forces and other notions. Although his mathematical theory worked beautifully, he wasn't able to provide a coherent physical justification and thought his construct only applied to electromagnetic theory. A correct interpretation of what he had found had to wait for Albert Einstein who used this to develop his theory of special relativity.

The Theory of Relativity

Einstein came along, suggesting that the phenomenon uncovered by Lorentz and others was not specific to electromagnetism but instead applied more broadly to physics in general. Among the consequences of these ideas, Einstein discovered a compact formula equating mass and energy that ranks among the most renowned in the history of science

$$E = mc^2$$

His theory tells us that notions of space and especially time, which had always been regarded as absolute, actually depend on the the velocity of the observer. Einstein discovered, moreover, that the Lorentz transformation is a physical transformation of space-time and not merely a mathematical trick needed to keep the Maxwell equations consistent. This proposition initially met with some resistance from the physics community, but its merits have since been irrefutably affirmed. Einstein's special theory of relativity involved linear transformations as one moves from a particular inertial frame of reference to another. As such, special relativity was mathematically rather simple and perhaps, to some palates, as far as its mathematical complexity, even a little boring using only elementary linear algebra. This helps illustrate the fact that deep physical ideas do not necessarily have to come from deep or complicated mathematics; they just need to come from *self-consistent* mathematics.

Einstein then went further, embarking on a reexamination of Newton's theory of gravitation. Georg Friedrich Bernhard Riemann had already introduced, a few decades earlier, a new form of geometry named after him. Riemannian geometry, as the approach was called, does not assume Euclid's fifth postulate and consequently allowed for phenomena such as triangles whose angles did not add up to 180° when the space in which they sit is curved (see Fig.4). Johann Carl Friedrich Gauss, who had been Riemann's professor, had previously suspected that such phenomena may occur in the real world and be measurable. Quite remarkably, it is said that Gauss proposed that our universe was curved. It is not clear whether this is a legend or accurate, but according to one account he tried to measure the curvature of space by measuring three angles of a triangle, where the vertices were the summits of mountains (see Fig.5), to see if they added up to 180°, assuming that the light rays are straight lines and form edges of the triangle. His measurements showed that the three angles did add up to 180°, within the bounds of experimental error, indicating that even if our universe is curved, that curvature was too small for him to discern.

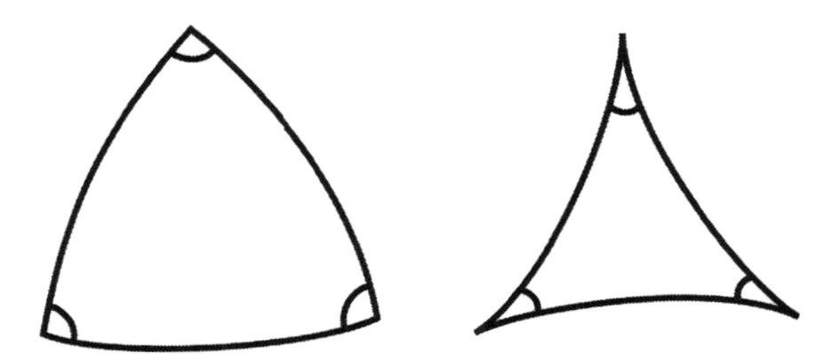

FIGURE 4. Non-Euclidean geometries: Euclid's parallel postulate is not assumed, so triangles do not necessarily have angles that add up to 180°

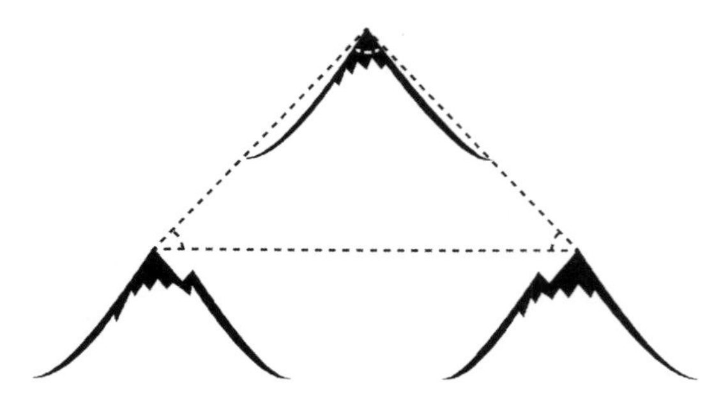

FIGURE 5. According to a legend (whose validity is disputed), Gauss tried to measure the curvature of space by measuring the interior angles formed by a triangle whose vertices are the tops of three mountain peaks. An interesting idea though no out-of-the ordinary curvature showed up in this exercise.

It is perhaps not surprising that Riemann also thought that Riemannian geometry ought to have some physical applications, and he even speculated that it might be used to unify the theory of electricity and magnetism with that of gravity. The application of Riemannian geometry to physics, however, had to wait for Einstein's reformulation of Newtonian gravity and his discovery of the analog of Maxwell's equations for gravity, the general theory of relativity. In general relativity, a completely geometric theory of gravity, the paths of free-falling objects in a gravitation field were simply straight lines (or geodesics) in the curved geometry of space and time. They appear "curved" (as if accelerating) because space itself is curved, just as the shortest distance between two points on the surface of an orange–the analog of a straight line–is curved too.

Today we know, based on Einstein's well-tested theory, that the universe is indeed curved. We also know that Gauss was on the right track, but he was trying to measure a curvature that was too small for him to detect. Riemann and Gauss were mathematicians, but some of the interesting mathematics they discovered later found its way into physics through Einstein's general theory of relativity. Here we see another example of the mutual help that physics and mathematics have given each other, advancing both fields in the process. Unlike special relativity, where the math involved was almost trivial, the mathematics of general relativity was very complicated and deep. Radical as these ideas were, the almost simultaneous advent of quantum mechanics was even more mysterious and baffling to scientists, even to Einstein–someone who was never fully comfortable with the field he helped pioneer.

Quantum Mechanics

Quantum mechanics introduced the strange notion of grounding physics in *probability*. For many physicists, this represented a step backwards since it implied that we could no longer predict for sure how nature would behave. Physical systems were subject to random fluctuations, meaning that chance, rather than certainty, was the rule of the day. It was for this reason that Einstein raised his oft-repeated objection to quantum mechanics: "God does not play dice with the universe." Quantum mechanics is also heavily counterintuitive, even to modern physicists and some of its leading practitioners. As Richard Feynman famously declared, "Anyone who says that they understand quantum mechanics is lying!" Nevertheless, the physics community has long embraced quantum mechanics for the simple reason that it agrees with experiments amazingly well.

The clash between quantum mechanics and established principles of physics led to some interesting conundrums. In the 1920s, physicists noticed that electrons seemed to possess an additional "degree of freedom"–an independent, defining feature that they called spin. Though it was similar to the conventional meaning of the term, it also had striking differences.

Erwin Schrödinger already had an equation that described quantum mechanics for speeds much less than the speed of light (the Schrödinger equation), but Dirac wanted to combine special relativity with quantum mechanics to account for speeds close to that of light. In so doing, Dirac found that he *needed* an extra degree of freedom, thus explaining the origin of this new kind of spin. Math stepped in again, to reconcile two areas of physics and this, as I will now show, has opened up new avenues in physics.

To understand the situation better, let's first take a look at the non-relativistic Schrödinger equation[4]:

$$\hat{E} = \frac{\hat{p}^2}{2m} + \hat{V}.$$

However, Dirac wanted an equation that was consistent with, and of the same form as, Einstein's well-known equation from special relativity:

$$\hat{E}^2 = \hat{p}^2 c^2 + m^2 c^4$$

To get a result that was also analogous to the Schrödinger equation, Dirac wanted to reduce the power of E in the above equation from E^2 to E without literally taking a square root. He realized that 4 x 4 matrices were needed to define his equation, and he introduced four matrices, α_k and β, such that:

$$\hat{E} = \sum_{k=1}^{3} \alpha_k p_k c + \beta m c^2,$$

With suitable choices of these matrices, it turned out that the square of this relation gives rise to Einstein's relation. The degrees of freedom of the electron's spin, moreover, arose from the same matrices. A mathematical idea thus led Dirac to a successful explanation of the origin of electron spin, again illustrating how abstract math can shed light on physics. Dirac's

[4]The terms in this equation are operators. $\hat{E}$ is the energy operator $i\hbar\frac{\partial}{\partial t}$, and $\hat{p}$ is the momentum operator $-i\hbar\nabla$ (mass m is still a number). $\hat{V}$ is the potential energy operator.

equation is one of the most celebrated statements not only in physics but also in mathematics and has been studied ever since by researchers from both disciplines.

Wolfgang Pauli, however, soon pointed out to Dirac that his equation allows for arbitrarily *negative* energy states. Dirac accepted this as a major problem to be solved.[5] He tried to do away with this problem by using Pauli's exclusion principle, which states that no two electrons can share the same orbit. Dirac suggested that the orbits corresponding to negative energies were already occupied (see Fig.6). The group of particles with negative energy are referred to as the "Dirac sea." Therefore, no other electron can be put in the negative energy orbits, thus solving the problem!

Physicists noticed, however, that this idea raised the strange possibility that particles could be knocked out of the sea to a higher energy state, thus leaving behind a hole of positive charge with the same magnitude as the electron's charge, behaving as a new particle with a charge opposite that of an electron. Dirac tried to dismiss the issue at first, claiming that the new particle with positive charge was nothing but a proton. But other physicists noted that the positively charged particle needed to have the same mass as the electron, as follows from Dirac's equation,

[5]For a very readable account of this see "*The Strangest Man: The Hidden Life of Paul Dirac, Mystic of the Atom*" by Graham Farmelo.

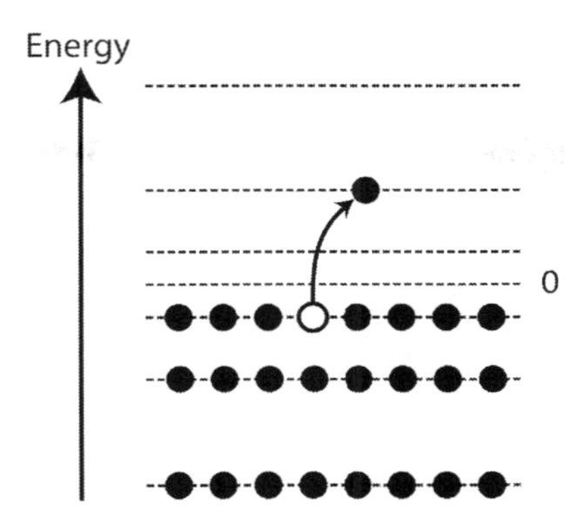

FIGURE 6. Dirac's equation yielded solutions with both positive and negative energies. Dirac tried to explain this away by arguing that the negative energy states are filled with a "Dirac sea" of electrons. When one of the electrons from the "sea" jumps to a positive energy state, it leaves behind a hole–a particle with positive charge that is otherwise identical to the electron.

but the proton is about 2000 times heavier than the electron. Finally Dirac had to accept the fact that his equation demanded the existence of a positively charged particle with the mass of an electron–which was later called the *anti-particle* counterpart of the electron. As no such particle was known to exist, his theory started coming under serious doubts. Even Dirac began to talk less about this aspect of his equation until, shortly thereafter, Carl Anderson found experimental evidence for this particle, dubbed the positron, from an examination of cosmic rays in his particle chamber. Sure enough, the positron had exactly the same properties as the electron, except for its opposite electric charge. Once again, mathematical elegance led to the prediction of new physics, which was hard to believe at first yet was eventually shown to be true.

Quantum mechanics initially had a limited domain of applicability, and it needed to be reformulated to be applicable to Maxwell's field theory of electric and magnetic forces. This reformulation was done by Richard Feynman and others, borrowing heavily from the approach taken by Euler and Lagrange in their reformulations of Newtonian mechanics, which we now turn to.

Quantum Field Theory

The classical picture, as we have discussed, holds that particles take paths that minimize the action. Quantum field theory presents a more complicated view in which a particle doesn't just take one path; it takes all possible paths, and a phase (a complex number of unit length) is assigned to each path. The probability of a particle going from any initial point to any final point is proportional to the sum of the phases. We will now put this into more technical language (that some readers might choose to skip over).

Feynman's *path integral* reformulation of quantum mechanics, as applied to particles, postulates that between points (x_1, t_1)

and (x_2, t_2), a particle follows paths weighted by the exponential of the *action*:

$$\int \mathcal{D}(X(t))e^{\left(\frac{i}{\hbar}\int(K-V)\,dt\right)}$$

where $\hbar$ is Planck's constant and the integral is over the space of all paths between (x_1, t_1) and (x_2, t_2). The integral is a complex number whose modulus squared gives the probability of a particle going from x_1 at time t_1 to x_2 at time t_2. The classical paths correspond to when $\hbar \to 0$. In this limit, the stationary phase method for evaluating the integral gives, as a good approximation, the paths that are extremum with respect to variations of the path. In particular, they correspond to the classical trajectories. Feynman's reformulation of Schrödinger's quantum mechanics produced a theory that is similar to the Euler-Lagrange reformulation of Newtonian mechanics as an action principle. The Euler-Lagrange formulation of Newtonian mechanics was easily adaptable to quantum mechanics (unlike the original formulation of Newton's laws), which is why we view them as more fundamental nowadays.

Notice that the mathematical reformulation of quantum mechanics in terms of path-integrals involved an infinite dimensional integral, because the space of all possible paths is infinite dimensional. Nevertheless, this has been made precise mathematically. However, Feynman also applied this path-integral approach to Maxwell's theory of electromagnetism, integrating over all electric and magnetic fields. This includes integration over the infinite-dimensional space of functions on $\mathbb{R}^4$. The mathematical complexity of this is far more than integration over the infinite dimensional space of paths.

This is a primary focus of quantum field theory, the mathematical underpinning of which is still being developed some 70 years after its original introduction! Despite not having a mathematically rigorous formulation of quantum field theories, physicists have developed an array of computational tools, including various approximation techniques, whose results match with experimental results to fantastic accuracy.

Quantum Gravity

In their early attempts to build upon Feynman's theory, physicists were not able to reconcile the general theory of relativity with quantum field theory in a unified theory of *quantum gravity* that described gravity at the level of individual particles. Using computational techniques developed for quantum field theories, one finds that the probabilities of physical amplitudes involving quantum aspects of gravity–such as the scattering of two quanta of gravitational waves hitting one another (what is called two "gravitons")–can yield infinitely large numbers. That's a serious problem, because a probability greater than one–let alone infinitely big–is a meaningless concept.

It should also be noted that even if we did have a consistent quantum theory of gravity, confirmation of such a theory would be well beyond current experimental means, as it would require probing energies many, many orders of magnitude above anything that could conceivably be produced in a laboratory setting. Some physicists were hesitant to work on quantum gravity theories, given the unlikelihood of empirical validation and the fact that attempts to fuse quantum mechanics with general relativity seemed to lead to nonsensical results. However, in light of the examples of Maxwell, Dirac, and others, many physicists know that, far from being a headache, an apparent contradiction can be a gift–a chance for a breakthrough–whose reconciliation has often moved physics forward. And that is why physicists have kept on trying to solve this inconsistency in the hopes of constructing a workable, unified theory.

A possible solution has emerged from an unexpected quarter. In the late 1960s, physicists were puzzled by the results of experiments involving the scattering of subatomic particles called hadrons. They looked at two kinds of processes. In the first case, one particle emitted something that the second particle absorbed. In the second case, two particles merged to form one particle before dividing again into two particles (see Fig.7). Although these processes seemed quite different, they led to the same outcomes. Physicists didn't know why this was this case, although they presumed that a new symmetry was involved.

Researchers subsequently discovered that this symmetry could be realized–and the two seemingly different physical processes could be seen as one and the same–if the point-like particles in their original models were replaced by more elongated, vibrating objects called strings. Strings were initially conjured up as mathematical objects that lacked physical justification, but the idea has since taken hold and proven to be quite fruitful. This was, in fact, the genesis of *string theory*–a theory in which particles (hadrons) were replaced by strings as the basic building blocks of nature. In this way, the above symmetry came to have a geometric interpretation: As strings move, they form tubes, and as they join and split, they create surfaces. The two scattering channels correspond to the same diagram for strings, which thus explained this symmetry (see Fig.7).

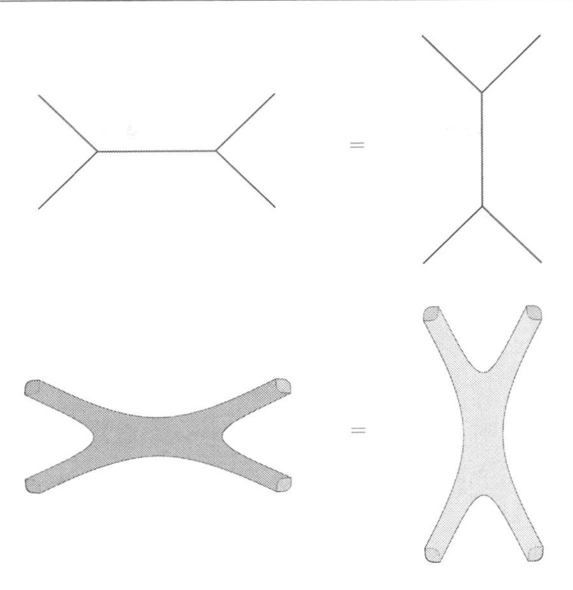

FIGURE 7. String scattering diagram for the two inequivalent particle scattering processes has the same topology

It was later found that strings were not a good description of hadrons, though they were a good description of quantum aspects of gravity. The lowest-energy state had properties (a massless particle of spin 2) that agreed with those of the graviton, which is the quantum excitation of gravity (just as the photon is the quantum excitation of electromagnetism), making it a prime candidate for a quantum theory of gravity. If we think about gravitons as tiny strings instead of point particles, many infinities that had been plaguing earlier theories disappeared.

String theory involves a great deal of modern math. The field has been strongly influenced, and indeed shaped by math; in return, physics in the form of string theory has had a great impact on pure mathematics. String theory is currently viewed as the prime candidate to describe a quantum theory of gravity. Moreover, as a byproduct, it also seems to unify all the other forces into a single framework, with all forces being manifestations of strings and their splitting and joining. This is where fundamental physics is at the moment: We have a theory that is very rich mathematically (as we will discuss later) but has not yet been verified experimentally–and probably will not be in the near future due to the extremely small size of the strings.

2. Symmetry and Conservation Laws

Symmetry is alluring to the eye and perhaps reassuring to the senses, as if reinforcing the notion that there is some underlying (or overarching) structure and order to our world. People have long marveled at the intricacy of hexagonallyshaped snowflakes, an often-cited example of nature's beauty. On a similar note, human faces that appear symmetrical are generally regarded as more attractive. Symmetry has also had a profound and rather obvious influence on architecture, which can be dramatically seen, for instance, from a straight-on view of the Taj Mahal, one of the Seven Wonders of the modern world.

Symmetry also plays a surprisingly powerful role in the laws and workings of physics, a role that goes far beyond matters of aesthetic appeal. To a physicist, symmetry is not merely a property of an object–such as a regular hexagon or octagon, that reflects a perfectly balanced construction and inner constitution. Symmetry often means an operation performed on an object or system that leaves it completely unchanged. Operations of this sort include rotations of an equilateral triangle about its center by 120 degrees or rotations of a square by 90 degrees. Rotations of a circle or sphere about its center are examples of continuous symmetries, valid for any and all degrees. A regular pentagon, on the other hand, has discrete symmetry: rotations of 72 degrees, or multiples thereof, leave it unchanged.

Yet the sweep of symmetry extends much further still, a fact illustrated by a theorem proven more than a century ago by Emmy Noether, a German mathematician. Noether showed that for every continuous symmetry of nature there is a corresponding conservation law. Using this theorem, one can derive important principles of physics–such as the conservation of energy, linear momentum, and angular momentum–strictly from mathematical arguments stemming from symmetry, as will be discussed later in this chapter.

Motivating Puzzles

Puzzles, as we have said before, can be effective tools for revealing the subtle interplay between physics and math. And they can be particularly useful in illustrating the link between symmetry and conservation laws, as the following examples hopefully reveal.

Puzzle

Our assignment, should we choose to accept, is to cover a standard chessboard (of 64 squares) with dominoes. Each domino covers two adjacent squares. The catch is that we have only 31 pieces of domino, which will leave two squares uncovered (Fig.8). Can we arrange the dominoes to cover all the squares except for the two diagonally opposite corners of the board?

FIGURE 8. Covering chess board with 31 dominoes covering all squares except for two diagonal corners.

Solution

When you put a domino on a chessboard, it covers a black square and a white square. Since the opposite corners have the same color, this problem cannot be solved. This is an example of a conservation law. Let N_{black} and N_{white} denote the number of black and white squares that are not covered. As we place the dominoes on the board, the numbers N_{black} and N_{white} change, but the quantity $\Delta = N_{black} - N_{white}$ does not change because each domino covers exactly one black and one white square. Δ, in other words, is a conserved quantity; it remains constant and unvarying in time.

Supposing we had succeeded in performing the task as requested, we would have ended up with $\Delta = 2$. But we started, before laying down any dominoes, with $\Delta = 32 - 32 = 0$, and this number, being a conserved quantity, could not have changed, which leads to a contradiction. In other words, there is no solution to the problem as originally posed.

Puzzle

Suppose there is a 4×6 grid with an entrance on the upper righthand corner and an exit in the leftmost square of the second row.

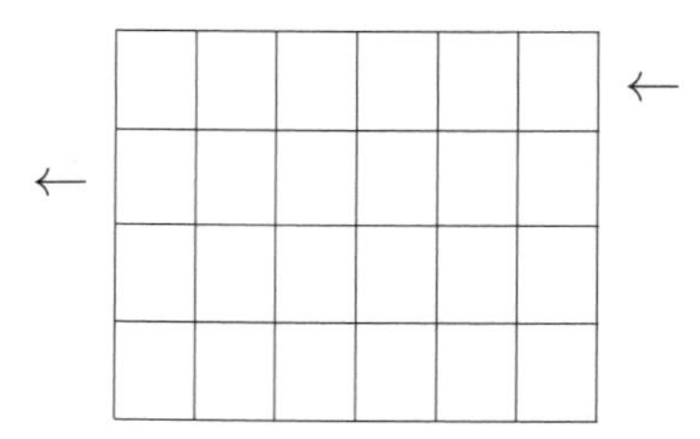

Is it possible to traverse this grid, moving only horizontally and vertically, and visiting each square exactly once?

Solution

Unfortunately, for those seeking a positive answer, it is not possible. Consider coloring each square alternating between ■ and □ with no two adjacent squares colored in the same way. Then the entrance and exit squares are the *same* color, but each step goes from a square of one coloring to another. So the last step, which is an even-numbered step, should have been black. Therefore, it is impossible to start and end on the same color square, while landing on each square exactly once.

So we see that symmetry, in this case the symmetry between differently colored squares, is a formidable invariant. However, coloring does not completely dictate feasibility in this entrance-exit grid problem. Consider the following example, with the exit and the entrance being the leftmost and rightmost squares on the second row.

We cannot disprove the feasibility of this entrance-exit grid problem outright, using mod 2 (even number-odd number) arguments alone, but it becomes clear upon trying the task that it cannot be achieved.

Symmetry

Symmetry, as we have stated, involves a transformation of a system that leaves it looking the same as it did before. As the figure below illustrates (Fig.9), an isosceles triangle has a symmetry of reflection about one of its altitudes.

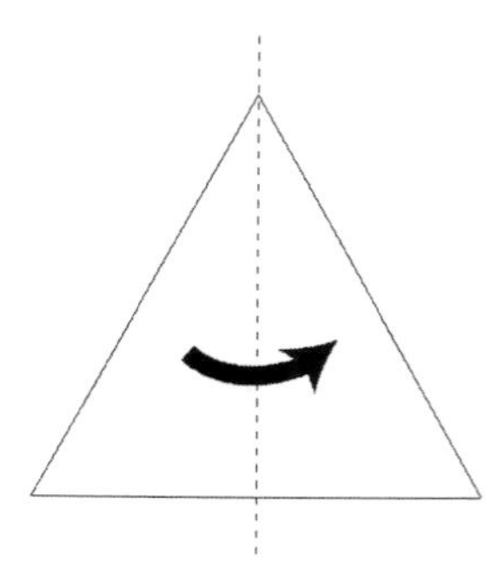

FIGURE 9. Isosceles triangles enjoy a reflection symmetry about one of their altitudes

Symmetries are closely related to the concept of invariances. This term, which can be used almost interchangeably with symmetries, refers to performing some operation that leaves the original object or configuration untouched or invariant.

Symmetry is a pervasive principle in physics, and we may think that symmetries should automatically be built into natural laws as intrinsic, inviolable features. But there are some caveats. For example, we might imagine that physical laws are invariant under symmetries such as reflections. In particular, if something is taking place one way, we might assume that its mirror image should also be physically possible. While in some cases this is true, in other cases it is demonstrably false. For example, there are particles that have a handedness (meaning that their spin and direction of motion obey a right-hand rule). But the mirror reflection of this particle would have an opposite handedness, and the opposite-handed particle either does not exist or has different properties. For example, electrons have a handedness in that electrons which spin clockwise versus anti-clockwise relative to the direction of their motion have different interactions with other particles. Physics, therefore, is not invariant under reflection. Reflection symmetry is often referred to as "parity."

Puzzle

Consider the hearts in a deck of cards. How can you arrange them such that if we alternate between flipping the top card and putting it on the table, and putting the next card underneath the deck (starting with putting the first card underneath), then the cards come out in exactly the order of $1, 2, \ldots, J, Q, K$?

Solution

There are many solutions, but a very simple approach is the following: lay out the cards in order–$1, 2, \ldots, J, Q, K$–and simply invert the operations as if you are playing the movie backwards (alternating between taking a card from the table and placing it on the top of the deck and taking the card at the bottom of the deck and putting it at the top) until the deck has been reconstructed.

What does this puzzle have to do with symmetry? This is an example of a *time-reversal* operation which is the temporal version of reflection symmetry in space we already discussed. Time reversal is a symmetry in some physical systems. In the present context, even if we reverse the arrow of time and things go backwards, we would get the card deck arranged such that when we roll time forward, it unwinds exactly the way we want.

Another symmetry we briefly discussed in the last chapter is the symmetry between matter and anti-matter (such as an electron and a positron), whose only difference is the sign of their respective charges. This symmetry is referred to as "charge conjugation." It turns out that time-reversal, parity (reflection in space), and charge conjugation are not actual physical symmetries of nature when taken alone. But it is a deep fact that combining Einstein's theory of relativity and quantum mechanics leads to the statement that when these three features are taken together as one, they do represent a symmetry in physics. In other words, if we start with any physical system and consider its mirror image, and reverse the arrow of time making it go backward, and replace every particle with its anti-particle we get a viable physical system.[6]

[6] One could argue that time reversal symmetry is not an exact symmetry in physics because we have the Big Bang that marks a definite starting point and, hence, a clear direction in time. However, that statement, by itself, does not tell us that time reversal is not a symmetry, because the time

Physics does have more powerful *continuous* symmetries. There are, for instance, *translations.* Consider a straight line. It has a translational symmetry (Fig.10):

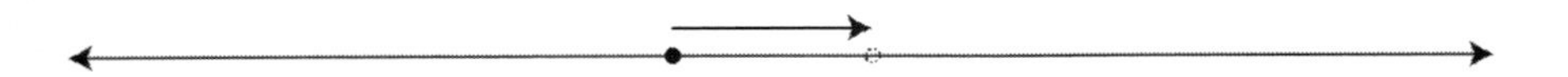

FIGURE 10. Moving the line along itself leads back to the line.

If we move the line along itself, it comes back to itself. It is, in other words, invariant. Of course, for this to be a symmetry of an actual physical system, we would need to move everything on the line in order for the physics to look the same. Translation in time is another symmetry. If I do an experiment today and do the same experiment tomorrow (assuming that everything in the universe is the same as today), then the experiments and their outcomes will be the same.

Another group of continuous symmetries is afforded by *rotation,* best illustrated by a sphere, which is invariant under all rotations around any of its axes (Fig.11).

---

reversed version of that, i.e. a contracting universe, would also lead to a viable universe.

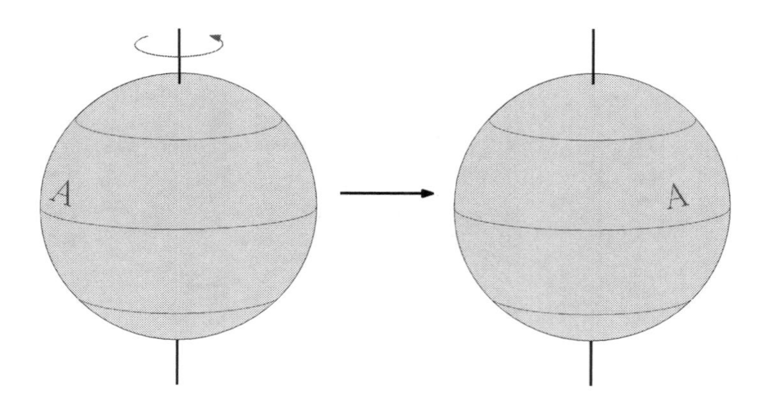

FIGURE 11. Rotating a sphere about its axis is a symmetry.

While most people are familiar with the notion of rotational symmetry in space, some of these symmetries can be rather subtle. Lorentz transformations, for example, are rotations in four-dimensional space-time, which literally involve the mixing of space and time: space is rotated in the time direction, while time is rotated in the space direction. Rotations of space and time coordinates into one another lead to a more generalized kind of symmetry that is specifically demanded by Einstein's theory of special relativity.[7]

[7] For mathematically more sophisticated readers: the group of rotations in 3-space is $SO(3)$ which leaves lengths $x^2 + y^2 + z^2$ invariant, and the Lorentz group of rotations in space-time is $SO(3,1)$ which includes transformations which mix space and time into one another such that $x^2 + y^2 + z^2 - c^2t^2$ does not change, where c is the speed of light.

Noether's Theorem

There is a deep relation between symmetries and conservation laws. By conservation we mean a quantity that stays constant and does not change over time. For example, if we have 10 indestructible balls that truly live up to that description, the number of balls will not change over time. It turns out that each symmetry in physics implies the existence of a quantity that is conserved in nature. That statement is a consequence of *Noether's Theorem*, published in 1918, which holds that for every continuous symmetry there must exist a corresponding conservation law. Apart from being beautiful, symmetries, as discussed earlier in this chapter, play an indispensable role in physics.

For example, the translation symmetry in space involves the idea that, all other things being equal, experiments performed in different locations should lead to identical results. Yet that same symmetry has even broader consequences: it inexorably leads to the conservation of momentum (mass times velocity). That, in itself, is rather amazing given that the conservation of momentum is a much more complicated statement than the obvious fact that the outcome of physics experiments don't depend on where we perform them.

These statements, in turn, can be used to reformulate Newtonian mechanics. Let us consider, for instance, the conservation of momentum for two particles, 1 and 2

$$\frac{d}{dt}(\vec{p_1} + \vec{p_2}) = 0$$

This differential equation shows that the change in momentum for the two particles is zero, meaning that total momentum is conserved–as it has to be, according to the laws of physics. We can then *define* the force on particle i to be

$$F_i = \frac{d}{dt}\vec{p_i}$$

In this way, we have recovered not only Newton's Second Law,

$$F = ma$$

but his Third Law as well: From the above equations, we can see that the sum of the forces on particle 1 and particle 2 is zero, which is another way of saying that these two forces are equal and opposite.

Remember when we talked about how it is not always clear as to which laws in physics are fundamental? The modern view of physics holds that the conservation of momentum is more fundamental than Newton's laws of motion because the former, though not the latter, is a direct consequence of a symmetry principle and that has a larger domain of applicability.

We are now starting to see the close connection between symmetry and invariance, or conservation. The three continuous symmetries we have discussed so far lead to the following conservations laws:

- Symmetry under space translation leads to the conservation of linear momentum.
- Symmetry under time translation leads to the conservation of energy.
- Symmetry under rotation leads to the conservation of angular momentum.

Puzzle.

We have two containers, one holding green paint and the other holding white paint. Assume that the containers are the same size and contain exactly the same amount of paint. Suppose we scoop out a small quantity of green paint in one cup and put it in the white paint container. Then we scoop out the same quantity of paint from the mixed container and put it back in the green container (Fig.12). What is higher? The concentration of green paint in the white paint container, or the concentration of white paint in the green paint container?

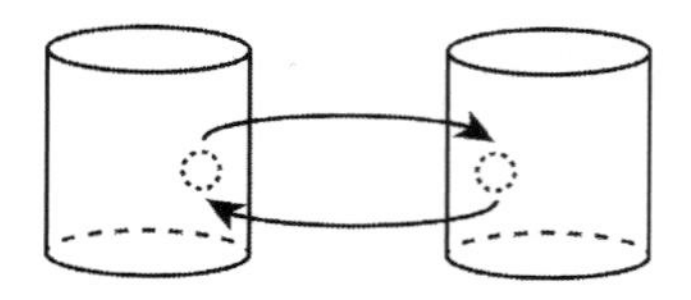

FIGURE 12. We scoop out a small quantity from the green paint container and put it in the white paint container and return the same amount after mixture.

Solution

The concentrations must end up equal! Since the total volumes were equal to begin with we end up with the same volume in each container at the end as well. Therefore, at the end of this process, any volume of green paint displaced from the green container can only be replaced by an equal volume of white paint, which follows from the conservation of the volume of the paints. Therefore, the amount of green paint missing from the green container is equal to the missing white paint from the white container. Thus we end up with the same concentrations for the mixture in both containers. This is a simple though nice illustration of the usefulness of conservation laws.

The idea behind this puzzle can also be illustrated with a deck of cards. Take 10 red cards and 10 black cards. Remove three red cards from the first group and mix it with the black cards. Then shuffle the black cards (which now contain three additional red cards) and draw three cards from that mix and put them back into the red stack. Convince yourself–using the general notion of conservation of red and black cards (as was the case for volumes in the previous example)–that there are, and have to be, as many black cards in the original red deck as there are red cards in the original black deck.

Puzzle

One numbered card is taken out from a deck. How can you quickly tell which numbered card was taken?

Solution

The units digit of the sum of the numbers on the numbered cards is 0, since all the numbered cards in the deck add up to 220. An efficient way of identifying the card in question is to add up the values of the remaining cards in mod 10 (i.e., keeping only the units digit). If the sum turns out to be 3, for instance, then you'll know right away that a 7 is missing; a sum of 9 would tell you that a 1 is missing. Another quick scan of the deck will reveal the suit of the taken card. The key principle at play here, once again, is to consider a conservation law, which in this case relates to the units digit of the sum of all cards.

This is somewhat reminiscent of one of the most famous examples of the application of conservation of energy to a pressing problem in physics–namely, Pauli's discovery of the elementary particle called a neutrino. Physicists had found that certain particle decay products seemed not to conserve energy. The energy of the particles coming out of the decay was not adding up to the energy of the original particle. So Pauli conjectured in 1930 that there must be a tiny new particle, which he called the *neutrino*, that was escaping detection and carrying off the extra energy, unnoticed. Pauli wagered that neutrinos would never be detected because they interacted so weakly with matter, but he lost that bet in 1956–the year neutrinos were discovered.

Puzzle

You have 10 boxes, each with 10 weights. Nine of the boxes have 1 kg weights, but one of them is defective and has a weight of $.9kg$. You have a digital scale that will display the total weight of any subset of the weights you choose. How can you detect the defective box in one weighing?

Solution

Label the boxes from 1 to 10, and take n weights from the nth box. You can determine which box is defective by the difference between the actual weight and the weight that would have been measured if none of the boxes were defective, which would have been $1 + 2 + \ldots + 10 = 55$.

Puzzle. Consider an infinite grid of squares in the first quadrant, or upper right-hand corner in standard Cartesian coordinates

(Fig.13).

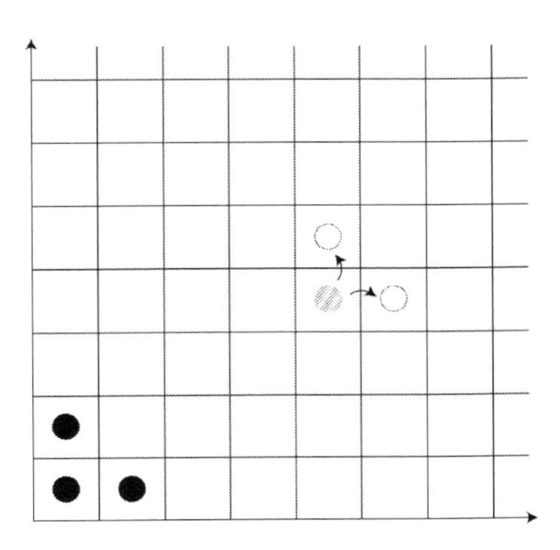

FIGURE 13. An infinite square grid with some pieces on the squares, where each piece can be replaced by two pieces–one directly above and one directly to its right–only if both squares are empty.

We will place pieces on some of the squares and allow them to change by a certain rule: Each piece can be replaced by two pieces, one on the square directly above and one on the square immediately to its right, only if both of those squares are unoccupied. Suppose we start with just three pieces on the first three squares in the lower left-hand corner of the grid, as shown in Fig.13. Your task is to use the operation discussed above in order to make sure that no pieces are left on these three squares. Is this possible?

Solution

No, this is not possible. To see why, let us assign a number to each square in this grid such that the numerical value of one square is equal to the sum of the two squares directly above and to its right. To be more specific, let us assign the number 1 to the first square in the lower left corner. Let us then assign the number 1/2 to the square above it and to the square immediately to the right. These two squares form what we will call the first diagonal. The next diagonal is made up of three squares to which each is assigned the number 1/4. The diagonal after that is made up of four squares, each assigned the number 1/8, and so forth. Each time we move out to the next diagonal, the number assigned is cut in half.

Now add the values of all the squares occupied by a piece. Note that the operation of replacing the pieces by the two pieces, one shifted horizontally to the right and one vertically up, preserves the total value. In other words, we have a *conservation law* of the total value of the squares occupied by the pieces. The value, to begin with, is $1 + 1/2 + 1/2 = 2$. Note that if all the board were occupied the total value would be summing over vertical columns first

$$= [1+(1/2)+(1/4)+(1/8)+...]\times[1+(1/2)+(1/4)+(1/8)+...] = 4.$$

Since the first three squares have a combined value of 2, it means that if you were to succeed in moving the pieces out of the first three squares, the total values they would occupy must still add up to 2. However, this can only be done if *all* the other squares in the grid get filled up because the total value of all the squares (including the first three) is 4. Hence, the task at hand is not possible in a finite number of steps or in a finite amount of time.

Puzzle

Let us suppose there is an odd number of soldiers in a field, all distinct distances from one another. All the soldiers are instructed to watch the soldier closest to them. Prove that at least one soldier is not watched.

Solution

Take the two soldiers closest together. They have no choice but to watch each other. Then pair off soldiers that are the next closest together, and so on. Given the odd number of soldiers, we will inevitably be left with one soldier who is not being watched.

But wait, could there be a fallacy in this argument? What if a soldier is closest to another one who has already been paired off? Let us say there are n unpaired soldiers, the first time this happens. Then one of these soldiers will not be watching any of the n soldiers because he is watching one of the soldiers who has already been paired off. Therefore, there are at most $n-1$

soldiers watching the n soldiers. At least one soldier, in other words, will be unwatched.

Puzzle

Can you draw the following shapes (Fig.14) without lifting your pen or retracing any segment?

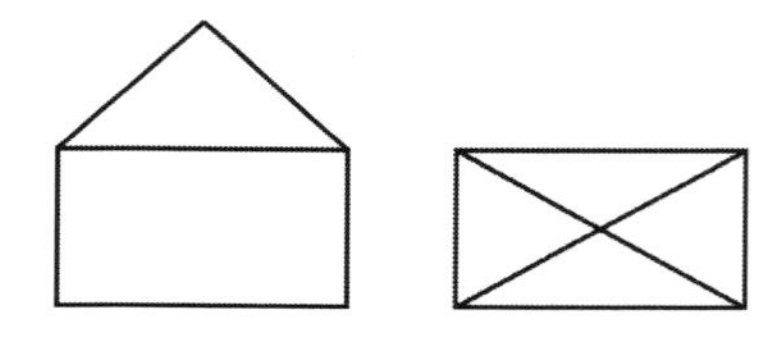

FIGURE 14. Trace the above shapes without lifting the pen or retracing the edges.

Solution

We can do so for the first shape but not the second. In order to draw a graph without lifting the pen, each vertex of the graph must have an even number of lines attached to it (except possibly for the first and last vertex, if they are distinct). This is because each intermediate vertex must be entered and exited the same number of times. An odd number of times would indicate that the vertex was the beginning or the end point rather than an intermediary, passing-through point. An equivalent though somewhat more abstract way of putting it is that the mod 2 number of edges on all intermediate vertices must be zero. Moreover, there can be at most two vertices that have odd numbers associated with them, and those can only be the vertices at the beginning and the end. It is thus clear for the first graph which vertices we can choose to start. The second diagram has 4 vertices with 3 edges attached to each so, according to what we have just said, it is not possible to trace that shape without lifting the pen.

Puzzle

There is a rectangular board on which you can place coins (Fig.15), subject to the following rules:

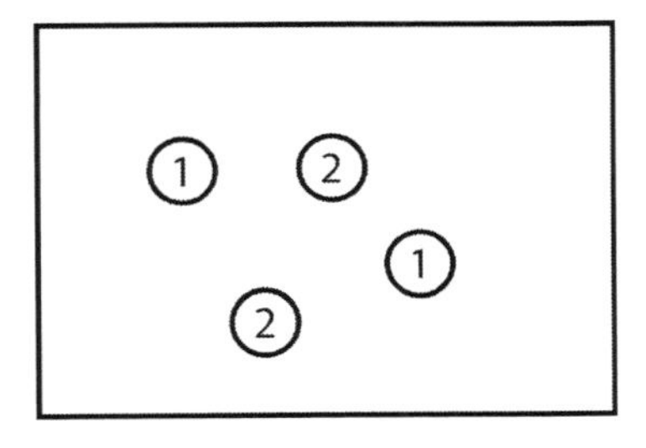

FIGURE 15. A game where two participants take turns placing non-overlapping coins on the table.

(1) The center of each coin must be completely contained within the boundaries of the table, and
(2) The coins cannot overlap.

Two players take turns placing coins. Assume we have an unlimited supply of coins! The last person able to place a coin under these rules wins. You play first. How can you be sure to win?

Solution

The winning move is to put the first coin at the center of the board (Fig.16).

Note that the board has order two symmetry, meaning that for any point on the table its reflection about the center of the table is also a point on the table. That means that no matter where the other person places the coin, you can always place one, in response, at the symmetrically opposite location. Owing to the symmetry of the board, if the other person's move is legal then so is yours.

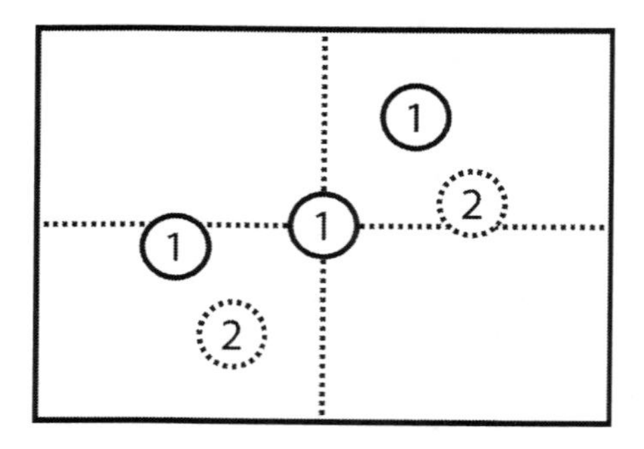

FIGURE 16. The winning strategy.

Puzzle

You are to play with two chess grandmasters simultaneously! The problem is that you don't know how to play chess very well. Nevertheless you wish to win at least one game or tie both games! On the first board the grandmaster is white. On the second board you are white. What should your strategy be?

Solution

The strategy is for you to mirror the moves on the other board. Whatever the grandmaster plays as white on the first board, you play on the second board, and whatever the second grandmaster responds with black pieces to your move, it will be your response with black pieces on the first board. In this way, by symmetry, the two games are identical and therefore their outcomes are identical. But you are on opposite sides in the two games. So if you lose on one side, you win on the other. Also if you tie on one of the games you tie on both.

Supersymmetry

There are more abstract symmetries in physics, and a prominent example of this is called supersymmetry. One consequence of this symmetry, assuming it is borne out in nature, is that every particle has a shadow particle called its superpartner. The superpartners would have identical properties to the original particle, except for having different spins. The selectron, for example, is the inferred superpartner of the electron. It has the same mass and charge as the electron but, unlike the electron, it has no spin.

On a more technical note, we can extend the dimension of space in theories with supersymmetry by adding additional coordinates. The new space is called a superspace. For example, we can have the superspace (x, y, z, t, θ). The extra coordinates, however, are rather different from other, more familiar coordinates: θ is an example of what is called a Grassmannian (or Fermionic) coordinate. Unlike the usual coordinates, in which we have $xy = yx$, a pair of Grassmann coordinates anti-commute. In other words: $\theta\alpha = -\alpha\theta$. Taking $\alpha = \theta$, we see that $\theta^2 = 0$. The Grassmann coordinates correspond to extra directions analogous to those of space, and the phenomenon of supersymmetry has to do with translational invariance in these directions. We stipulate the following conditions:

$$\theta \cdot \theta = 0 \tag{2.1}$$

$$\frac{\partial^2}{\partial \theta^2} = 0 \tag{2.2}$$

$$\theta \frac{\partial}{\partial \theta} = -\frac{\partial}{\partial \theta}\theta \tag{2.3}$$

Supersymmetry leads to another kind of symmetry that is the square root of translation in this space. For instance, let $f(x)$ be a function. If we change x by ϵ, then the value of the function changes by its derivative: $f(x+\epsilon) \approx f(x)+\epsilon f'(x)$. The operator $\frac{\partial}{\partial x}$ is the generator of the symmetry, in a sense that we will not make precise here. The square root of this generator of the translation symmetry would then be an operator D_θ such that $D_\theta^2 = \frac{\partial}{\partial x}$. This is difficult to imagine, but supersymmetry allows one to do precisely that. Consider the following:

$$D_\theta = \frac{\partial}{\partial \theta} + \theta \frac{\partial}{\partial x}$$

Then

$$D_\theta^2 = \frac{\partial^2}{\partial \theta^2} + \theta^2 \frac{\partial^2}{\partial x^2} + \frac{\partial}{\partial x} + \frac{\partial}{\partial x}\left(\theta \frac{\partial}{\partial \theta} + \frac{\partial}{\partial \theta}\theta\right)$$

If we remember our rules for anti-commuting variables, almost all terms vanish and D_θ^2 simplifies to $\frac{\partial}{\partial x}$. Note also that a function on this space would have a power series in θ of the form

$$f(x, \theta) = f(x) + \theta g(x) \tag{2.4}$$

There are no higher order terms because $\theta^2 = 0$! So a function on this superspace can be viewed as a pair of functions f and g. This is the analog of doubling one sees for each particle having a super partner.

The notion of supersymmetry may appear strange at first glance, but it is an essential ingredient of string theory, as well as in certain quantum field theories. Supersymmetry makes quantum mechanics look more classical by taming quantum fluctuations. However, there is currently no experimental evidence for the existence of selectrons (the supersymmetric partners of electrons) or of supersymmetry in general. Researchers are hoping that some evidence for it will emerge soon in collider experiments.

Quasi-crystals and Symmetry

Here is a somewhat unusual kind of symmetry, that of a *quasi-crystal.*

To set up this discussion, you have probably seen various tilings of the plane possessing symmetries by discrete groups: $\mathbb{Z}/2, \mathbb{Z}/3, \mathbb{Z}/4, \mathbb{Z}/6$ ($\mathbb{Z}/n$ refers to rotations of $2\pi/n$. The $\mathbb{Z}/2$ group, therefore, relates to symmetrical rotations of π radians or 180 degrees. $\mathbb{Z}/3$, similarly, relates to rotations of 120 degrees; $\mathbb{Z}/4$ relates to rotations of 90 degrees, and so forth).[8] See Fig.17.

[8]There is a somewhat simple mathematical reasoning showing that only these rotations show up as rotational symmetries of tilings on a plane: the matrices associated with rotations of order $2, 3, 4, 6$ are those with trace equal to an integer. Why should this be the case? Any transformation of a lattice $\simeq \mathbb{Z}^2$ can be written with integer entries because you can view the rotation as acting on some vectors which span the lattice which should therefore map vectors to integral combination of other vectors. Hence such matrices have integral trace. On the other hand, $\mathbb{Z}/5$ rotation would give a trace as $2cos(2\pi/5)$ which is not integer unlike $2cos(2\pi/n)$ for $n = 2, 3, 4, 6$ which is an integer.

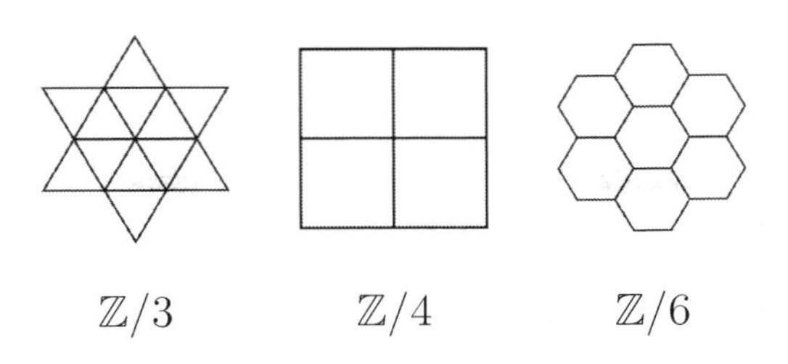

FIGURE 17. We can form symmetric, periodic patterns associated with 3-, 4-, and 6-fold symmetry.

There are also crystals possessing quasi-symmetries. These are known as quasi-crystals, which can be generated using *Penrose tilings*. The tiles are made up of five-sided objects that individually possess $\mathbb{Z}/5$ symmetry, but the crystal, as a whole, does not have rotational symmetry. It is quasi-crystal, because it is not strictly speaking periodic, but it is almost periodic (Fig.18). Quasi-crystals are objects like crystals that *almost* have symmetries, but not quite. Each local component looks symmetric, but there are no global symmetries, even though these structures are almost periodic.

Interestingly, a research fellow at Harvard, Peter Lu, and a physicist at Princeton University, Paul Steinhardt, found that many mosques built centuries ago are adorned by quasi-crystals, which shows that the general idea of Penrose tilings have been around roughly since the year 1200–long before Roger Penrose, for whom the tilings are named, started looking into them in the 1970s.[9]. So ancient civilizations also appreciated the beauty of the almost symmetric structure of quasi-crystals. But those architects had a different motivation from ours. They were not trying to model physics or experiment with symmetry principles; they were trying to create subtle but pleasing effects on the eye!

Many solids in nature are lattice-like, with symmetries of crystals as we have discussed. Interestingly, there are also compounds in nature that are instead quasi-crystals and enjoy their subtle symmetry patterns! Dan Shechtman was awarded the Nobel Prize in Chemistry in 2011 for the discovery of quasi-crystals in the natural world.[10]

FIGURE 18. Crystals with quasi-crystalline symmetries

[9] www.peterlu.org/sites/peterlu.org/files/BA.pdf

[10] See www.nobelprize.org/nobel_prizes/chemistry/laureates/2011/popular-chemistryprize2011.pdf

[11] Penrose figure made by Inductiveload on Wikimedia

Puzzle

What is EF in the following figure (here E is at the center of the square)?

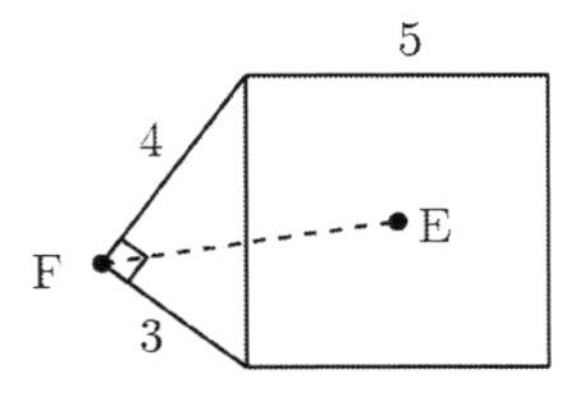

Solution

The key is to extend the figure to a symmetric one.

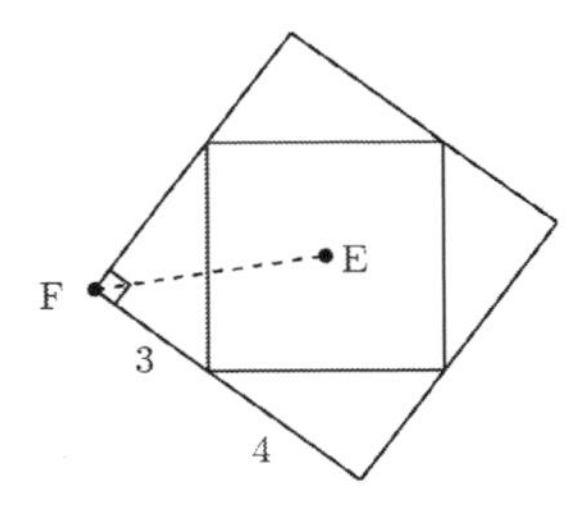

We now see that EF is half the diagonal of a square of side 7. Thus $EF = 7/\sqrt{2}$, which we know from the Pythagorean theorem. But we can work in the other direction too, as the above symmetrization is also one of the simplest ways to prove the Pythagorean theorem. With lengths of the sides of these right triangles being a and b (and the hypotenuse being c), the overall area of the bigger square is $(a + b)^2$, but this is made up of four triangles, each of area $(ab/2)$, and one square of area c^2. This leads to $(a + b)^2 - 4(ab/2) = a^2 + b^2 = c^2$–the most famous, and arguably the most important theorem in the history of geometry.

Strings and Conservation of Charge

The electric charge of an electron is -1 in units known as elementary charge. The electric charge of a proton is $+1$ in the same units. There are two basic properties of electric charge: All electric charges come in integer multiples of this unit and electric charge is conserved. What could be the explanation for both the discreteness of electric charge, as seen in nature, and the conservation law pertaining to charge?

In string theory, where particles are replaced by extended one-dimensional objects called strings, one often considers geometric situations such as the following: there is a loop (a *string*) on an infinite cylinder (Fig.19), where the circumference of the cylinder is viewed as an extra dimension (string theory enjoys more than 3-spatial dimensions–the extra ones are believed to be tiny, as we will discuss when we discuss dualities). Such a loop has a characteristic *winding number* that describes how many times it winds around the cylinder.

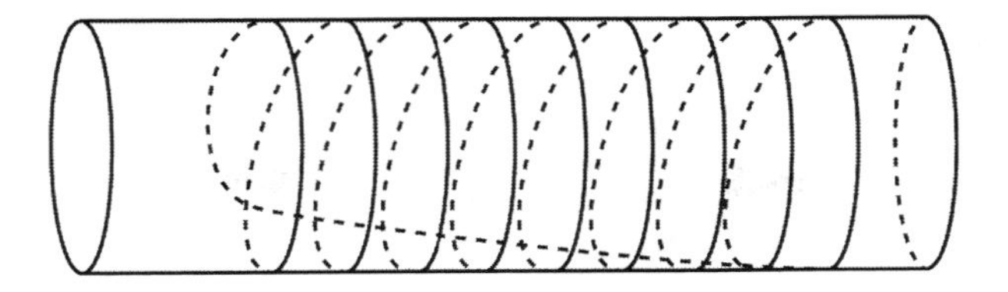

(A) The addition of multiple strings: $1+1+1+1+1+1+1+1+1 = 9$

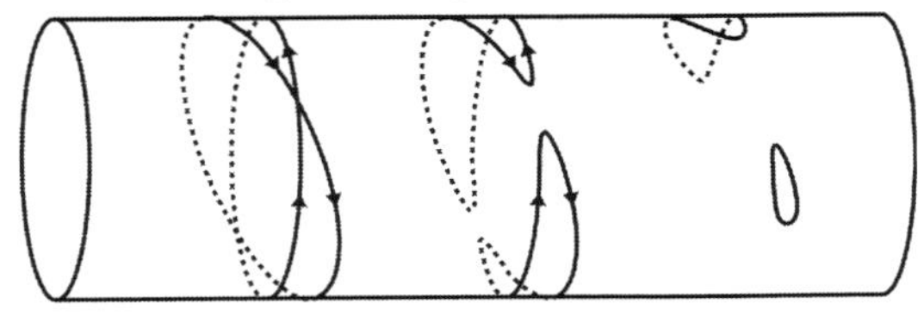

(B) The annihilation of two strings: $1 + (-1) = 0$

FIGURE 19. Charged particles as winding strings

This offers a possible explanation for the discreteness of electric charge. If charge is interpreted as the winding number of strings around a circle (or cylinder), then it must come in discrete multiples of a fundamental unit. What about the conservation of charge? The way in which two strings interact is by concatenation, meaning that, upon touching each other, two separate strings will reconnect. The addition of charge represented by concatenation of winding strings can become more subtle in some situations, as the next puzzle demonstrates.

Puzzle

Imagine you have picture frame hung by one string. Nailed to the wall are two pins. How can you wrap the string around the two nails (Fig.20) such that the picture does not fall, but as soon as either of the nails are removed the picture will fall? (as an extension, what about a situation with $N = 100$ pins, where the frame is hung but will again fall as soon as you remove any *one* of the nails?)

FIGURE 20. Hanging a frame with a wire on a wall with two pins.

Solution

In our previous example, involving the winding of a string, we had a conservation law that included an additive operation. This operation, like addition itself, is commutative and forms a so-called "abelian group." However, for strings winding around two different nails or centers, the order in which they wrap matters. In other words, the winding of strings around different centers leads to a "non-abelian" (i.e., non-commutative $gh \neq hg$) group, and the notion of a conservation law still exists, though it is more subtle than simply adding up the separate winding numbers.

To solve this puzzle, we will take advantage of this non-abelian feature.

Here is the basic idea: If we have a nail and wrap a string around it clockwise, we'll call it α; if we wrap a string around the nail in the counter-clockwise direction, we will call it α^{-1}. The product of α and α^{-1} is equal to 1, which means there is no net winding around the nail if we wrap the string sequentially around a nail clockwise and then counter-clockwise. The string is thus unwrapped. We do the same thing for the second nail, calling a winding in the clockwise direction β and a winding in the counter-clockwise direction β^{-1}. The non-commutativity comes into play if, for instance, you were to wind a string around the first nail and then the second nail and then unwind around the first and then unwind around the second (Fig.21). Unlike the previous case, now the picture will not come down because the order of operations matters.[12]

We can express some of these same ideas mathematically: The operation we just discussed can be represented by

$$[\alpha, \beta] = \alpha\beta\alpha^{-1}\beta^{-1}$$

[12]This problem can be generalized to N pins (including the case of $N = 100$) when the configuration of the winding string implements $[\alpha_N[\alpha_{N-1}[\cdots[\alpha_3[\alpha_2, \alpha_1]]]]$.

This element is considered non-trivial because the group is non-abelian. Non-triviality means that if we wind the string according to this prescription, it will not fall. Removing either nail is equivalent to making α or β equal to 1, in which case the product trivializes and the frame falls.

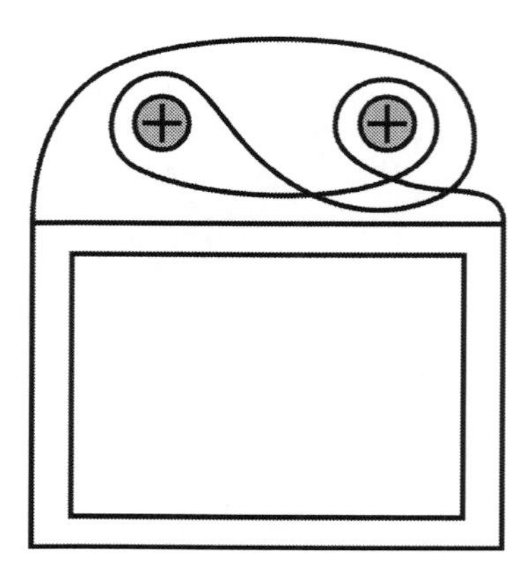

FIGURE 21. Solution for the hanging of the frame.

Spontaneous Symmetry Breaking

So far we have talked about symmetry and its important applications. The topic we will take up next is spontaneous symmetry breaking. In this context, the application of symmetry can lead to wildly unexpected results. Let us start out by saying what spontaneous symmetry breaking is *not*. You might suppose, for instance, that there is a symmetry breaking between up and down. Due to the presence of Earth's gravitational field, all directions are not the same, and indistinguishable from a physics standpoint, as they would be in a wholly symmetrical situation. But this is still not an example of *spontaneous* symmetry breaking because it is a consequence of the aforementioned gravitational field–a constant environmental condition, you might say, rather than a sudden natural shift that changes everything.

Spontaneous symmetry breaking, which we will discuss in the next chapter, is rather different. Moreover spontaneous symmetry breaking turns out to be a crucial phenomenon in physics. It explains why we exist, and why mass exists. Without mass, we would be zooming along at the speed of light!

3. Symmetry Breaking

In the last chapter, we illustrated some of the power of symmetries in solving puzzles, in studying physics and in the world and universe around us. We noted that symmetries are equivalent to conservation laws in physics and conservation laws, as you may have noticed, can be very useful. One basic application, as we saw in the puzzle with the cards, is that if things do not add up, we then know something is missing, and we can get information about the missing thing by counting what is, and is not, there. In this chapter we discuss the opposite concept: situations where symmetries are broken. And you may be surprised to find out that, in some cases, these broken symmetries can be more interesting and consequential in nature than unbroken ones.

One example is the asymmetry between matter and antimatter. The Big Bang, in principle, should have created equal amounts of matter and antimatter. If that situation had persisted, matter and antimatter particles would have eventually come into contact and annihilated each other in bursts of pure energy. But somehow the symmetry between matter and antimatter was broken by a tiny bit with one in a billion more of matter than antimatter, leaving behind a preponderance of matter after annihilation, to which we owe our existence!

Another example, less vital to our presence in the cosmos,

concerns a pencil balanced perfectly on its tip. It is an unstable configuration because the pencil will eventually fall. But while the pencil is upright, it's in a symmetric configuration, because it could fall any which way–no one direction is preferred or preordained. When the pencil eventually topples over, that symmetry–beautiful while it lasted–is then spontaneously broken. While a moment before, it might have fallen in any direction from 0 to 360 degrees, it has now singled out just one.

Here is a more mathematical example: Suppose we have a smooth real function of one variable, $f(x)$, and let us further suppose that it is an even function. In other words, f enjoys a reflection symmetry: $f(x) = f(-x)$. Our initial task is to find the critical points of f, i.e., points where $df/dx = 0$. Using symmetry we can immediately find one of the solutions

$$\left.\frac{df}{dx}\right|_x = \left.\frac{df(-x)}{dx}\right|_{-x} = -\left.\frac{df}{dx}\right|_{-x}$$

Therefore at $x = 0$, we have $\left.\frac{df}{dx}\right|_0 = -\left.\frac{df}{dx}\right|_0$. The only way for this to be true is $\left.\frac{df}{dx}\right|_0 = 0$.

Suppose we instead asked: Find a local *minimum* of $f(x)$ (assuming there is one). One's immediate guess may, on symmetry grounds, be $x = 0$. However, this is not necessarily so: One of two situations is possible as shown in Fig.22 and, depending on which of those it is, the local minimum may or may not be at $x = 0$.

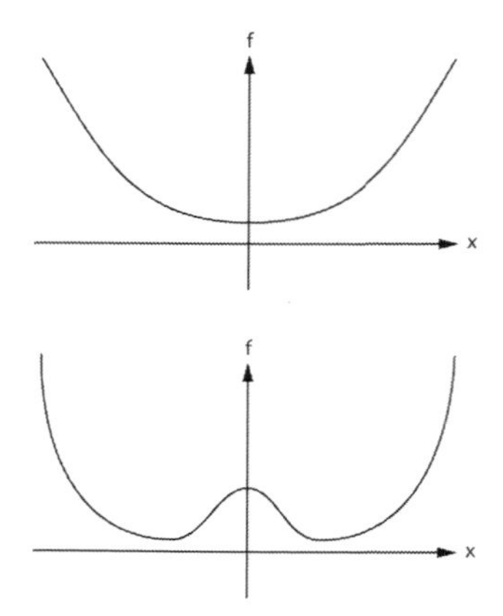

FIGURE 22. The minimum of an even function can preserve (as in the top figure) or break (as in the bottom one) the reflection symmetry.

If it is not at $x = 0$, we say the reflection symmetry is "*spontaneously broken.*" In other words, symmetry may mislead us in our thinking as to where the actual minimum should be. In the event that the symmetry is broken, there must be at least two minima.

Earth's Motion and Symmetry Breaking

Applications of the concept of symmetry to explain physical phenomena go all the way back to the ancient Greek philosophers, if not further. As we discussed in Chapter 1, the Greeks had already figured out that the Earth is a sphere. Moreover, they had known that it rotates about its axis, because all the stars at night seemed to revolve around the Northern star, Polaris, and they thought that would be rather unlikely and they postulated that instead the Earth is rotating but the stars are fixed. They also thought (incorrectly) that the center of the Earth is not moving because they thought if the Earth were moving the positions of stars would change, unlike what was seen night after night. The fact that the center of the Earth seemed stationary bothered them and they looked for an explanation of it. They knew about the rotational symmetry enjoyed by spheres. Based on symmetry considerations, their view of the heavens placed the Earth at the center of the universe. They then argued that since the Earth is at the center of the universe, there is no preferred direction in which to move: If it were to move, they reasoned, it would break the rotational symmetry. In order to preserve the rotational symmetry, they concluded, it better not move. This line of reasoning led to the picture that the center of the Earth is fixed at the center of the universe.

Aristotle challenged this argument. He argued that if a person (or even a donkey, for that matter) stands at the center of a circle with food evenly distributed around its perimeter, Fig.23, then it will eventually choose a direction to move and get to the circle or else it would starve![13] This movement, and in particular the choice of a direction to move, will necessarily

---

[13]see *On the Heavens* by Aristotle

break the circular symmetry that once prevailed. However, in real life and in the physical world, choices have to be made in symmetric situations that may lead to asymmetric results. Symmetry is, indeed, a wonderful thing–a source of great beauty–and magic in many ways. But is it a principle that should be put above everything else–a principle worth starving for? Aristotle brilliantly argued that symmetries are not preserved at all cost. The optimal choices are not always symmetric and they can spontaneously break!

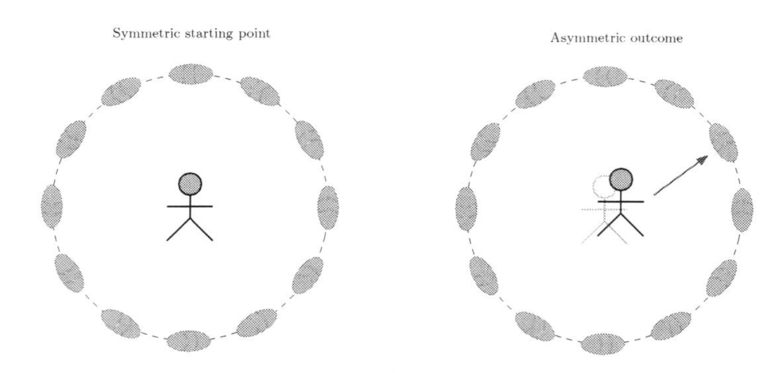

FIGURE 23. Aristotle gave the first example of spontaneous symmetry breaking: A person at the center of a circle, with loaves of bread equally distributed around its perimeter.

Spontaneous Symmetry Breaking

We are now going to delve into the notion of *spontaneous symmetry breaking.* The reason it is called "spontaneous" is that the starting point is a symmetric situation, yet the solution inexorably forces us to an asymmetric outcome. Let us return to the above example, which places us at the center of a circle, with food (say loaves of bread) evenly distributed about the perimeter. We could, of course, have broken symmetry by hand, for instance, by putting more food on one side of the circle than the other. In that case it would be clear as to what the preferred direction of motion would be–namely towards the point on the circle where more food is clustered. That would not be an example of spontaneous symmetry breaking, because the starting point was already asymmetric to begin with.

There are many other examples of symmetry breaking in nature. Evolution has shaped us, just as our environment has shaped evolution. We live on a planet where up and down are different due to gravity, which points in one direction, namely down. There is not an "up/down exchange symmetry," in other words, here on Earth things fall down; they do not fall up! Given that there is no symmetry between up and down, it therefore makes sense that our feet do not look anything like our head.

On the other hand, if we are standing on flat ground, everything in the plane is rotationally symmetric, yet evolution has broken that symmetry when it comes to human anatomy: our body does not enjoy horizontal circular symmetry. Our eyes, for example, point in specific directions rather than all around our heads. Nature has somehow found it to be more efficient, less wasteful of energy and other resources, to have eyes on one side of our bodies–facing "forward," as we put it. In the context of Aristotle's example, our eyes are in the front, so we can go towards the food! Even though our eyes are left-right symmetric, that is not the case for all of our body. For example, the human heart, for some reason, is located towards the left side of the thoracic cavity. The stomach too is more to the left, while the liver is mainly on the right.

It seems that even in nature, symmetry is not always the best solution. In modern physics, moreover, we are starting to see spontaneous symmetry breaking in many different contexts, and we continue to learn more about the important role it plays in this chapter.

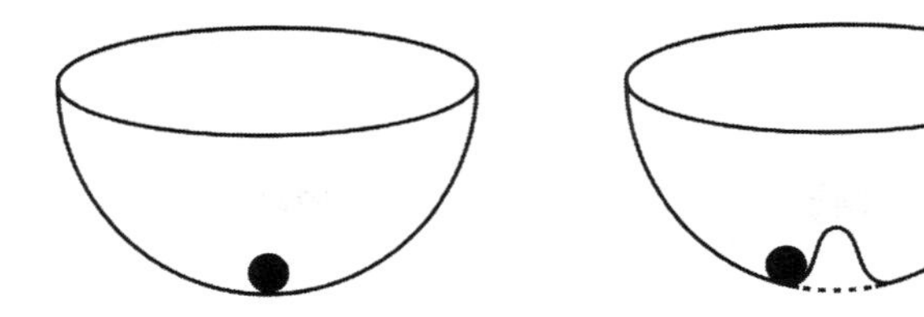

FIGURE 24. For a symmetric bowl with the bottom at the center, a ball rests at the center. This is not the case if the bottom of the bowl is not at the center.

Suppose there is a ball in a symmetric bowl, as in Fig.24 (like the bottom half of a sphere). Where will it settle? You can see that it should be at the very bottom, at the center of the bowl. One is tempted to say that this must be the case due to symmetry considerations. Suppose, however, that the bowl has a small bump at the very bottom, but the bump is centered in such a way that the bowl is still symmetric. Then there would be a whole family of locations at which the ball might naturally come to a rest *none* of which are at the center. In fact, symmetry demands that we have a whole ring of possibilities. This illustrates the fact that breaking symmetry, as a general matter, yields many solutions where there might previously have been just one. Notice that a small change in the conditions, Fig.25 (such as slightly tilting the bowl) will destroy the symmetry and move the solutions around considerably. Notice, also, that the asymmetric solutions are not frozen in the sense that a small "jiggle" will move them around. Putting that in other terms, if we were to break symmetry and tilt the bowl slightly, the ball can change its position drastically. So, this is an unstable situation, which is often the case when you break symmetry.

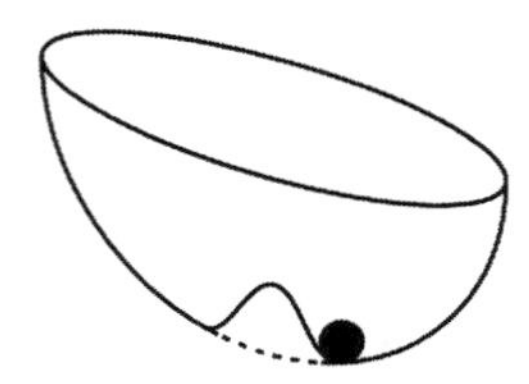

FIGURE 25. If the symmetry is broken by hand by a slight tilt of the bowl, a preferred bottom point will emerge where the ball settles.

Realization that there is a symmetry can depend on your perspective. A donkey on the boundary of a circle, close to where the food is, may not notice the symmetric shape, unlike a donkey at the center of the circle. The same is the case from the vantage point of a ball at the bottom of a bowl with a symmetric bump at the bottom Fig.26. One may be misled into thinking there is no symmetry if he or she happened to be situated in an asymmetric point.

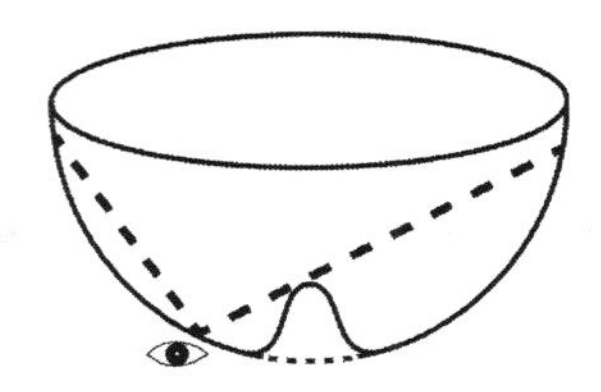

FIGURE 26. From the vantage point of a minimum at an asymmetric point, it is hard to appreciate the existence of rotational symmetry of the bowl.

The opposite can also be true: Sometimes we can break the symmetry by hand but symmetry can still be powerful enough to guide us toward solutions. For instance, suppose you want to find the center of mass of a rectangle. You can use calculus to find it, but it is easy to reason, drawing upon symmetry that the center of mass is, well, at the center. After this, you can go back and prove it rigorously. It is often useful in physics problems to move to the center-of-mass coordinates, because they have a special, built-in symmetry. However, what if the shape is not symmetric? Can you still use symmetry in some way to answer the same question?

Puzzle

What is the center of mass of Fig.27 (where the L shaped object is not assumed to have any symmetries)?

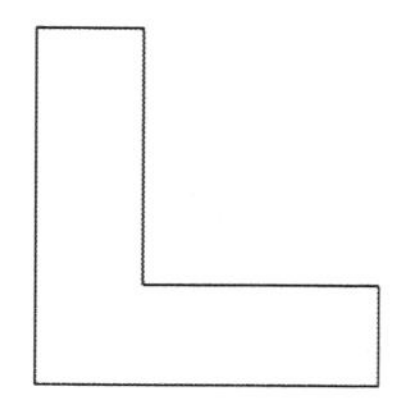

FIGURE 27. Can you find the center of mass of this asymmetric L-shaped object?

Solution

The center of mass lies on each of the dotted lines (connecting the two centers of rectangles where we divide the space in two different ways) as in Fig.28, hence it must be at their intersection. One lesson to draw here is that the symmetry principle is still powerful, even in situations that don't appear to be symmetric at all.

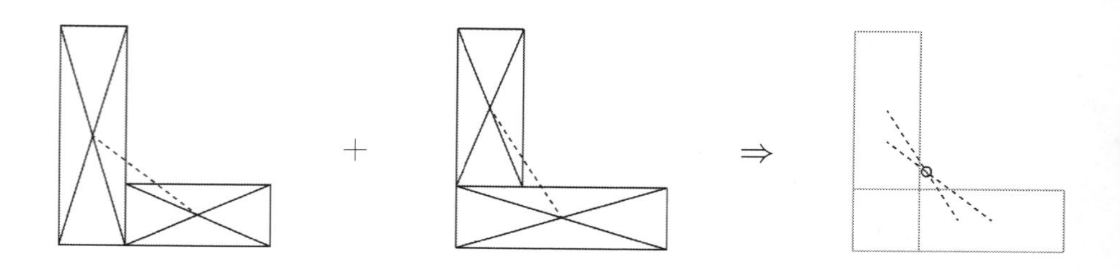

FIGURE 28. One can find the center of the L-shaped object by dividing it to two different pairs of rectangles.

By this method, we can find the center of mass of any quasi-rectangular shape without calculating lengths, even a complicated shape like this:

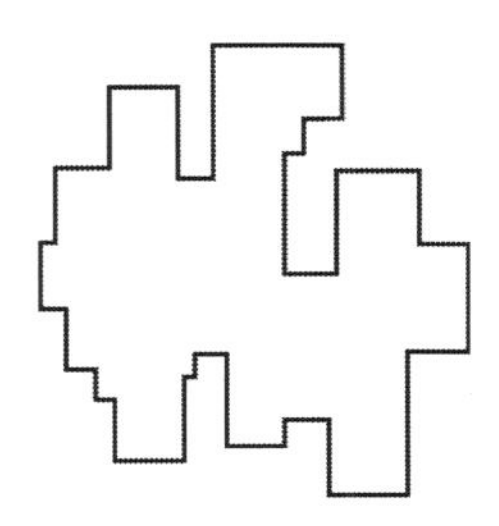

Spontaneous Symmetry Breaking and Magnets

Suppose we have a system of particles that has a degree of freedom called "spin," which can be either up or down. If you're familiar with chemistry, you can think of this as electron spin. Recall also our discussion in Chapter 1 of electron's spin and Dirac's explanation of it. Let us further suppose that when two particles are brought together, they "like" to have the same spin (both up or both down), in the sense that these states have lower energy overall. (Ferromagnetic materials, for example, act like magnets when electron spins are lined up in the same direction. But these materials lose their magnetic properties if the electron spins are randomly aligned, in which case the magnetic effects cancel out). Same-spin configurations have energy $E(\uparrow\uparrow) < E(\uparrow\downarrow)$, the former being the energy of same-spin and the latter being the energy of opposite-spin configurations. Imagine a lattice filled with such particles situated on a plane (Fig.29). That is the model we are considering here, which is called the Ising model.

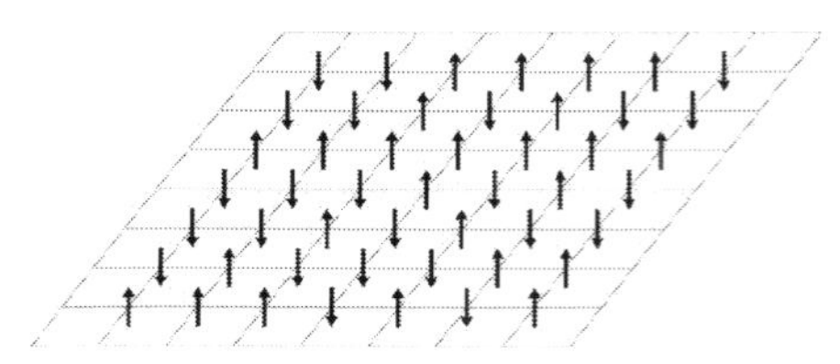

FIGURE 29. The Ising model involves up and down spins, where nearby spins prefer to align, to lower the total energy in the system.

For each pair of neighbors, there is some energy which we add up to get the total energy of the system. There is a clear minimum total energy configuration in which all spins are pointing in the same direction. If left alone, without any external influences, the lattice will find the minimum-energy state, as with the example of the marble in the bowl. Namely, all the spins will point up or all will point down.

However, suppose the lattice is in a heat reservoir that keeps the particles at a given temperature, which allows having configurations of spins which are not minimal in energy. To be precise, we assume that the probability that the system has total energy E is $p(E)$, and this is related to temperature according to "Boltzmann's rule," i.e., $p(E) \propto e^{-E/kT}$ (here k is the "Boltzmann constant." The lowest energy configuration has the highest probability, but for any positive temperature, any state is possible with some probability.

Let us count the spins as $+1$ or -1 for up or down. Let S denote the average spin, i.e., the sum of all the spins of the system divided by the total number of particles. What is S?

It must be zero! For any given configuration of spins that you might have, the opposite configuration (obtained by flipping the signs of all of the spins) has the *same* energy and thus the same probability to occur. That means that S equals $-S$, which is another way of saying that S equals zero.

So we see that the $\mathbb{Z}/2$ symmetry of the system passes on to S, forcing it to be zero. Is this consistent with our model of magnets? Magnetization results, as mentioned before, from an alignment of spins and the strength of it is proportional to S. But we just argued that the average spin must be 0. So why do magnets exist? We will now explain how this happens.

At the extreme low temperature – $T \to 0$ – the system settles to its absolute lowest energy state. This happens when all the spins point up or all the spins point down. So there are two ground states. Which one will the system settle in? That depends on where you start. For example, if you apply a small magnetic field, which forces the spins to align in one direction, then you can pick one of these two lowest energy configurations. Moreover, even after you turn off the magnetic field, you will stay in the same state. The reason is that to go from one of the two states to the other, you will have to flip *all* spins, and even though the final state would have the same energy, to get there the energy barrier for this is large. In this way, we end up at a small enough temperature with a phase that automatically picks out one direction of the spin. In other words, the $S \to -S$ symmetry of the problem has been spontaneously broken. The breaking of this symmetry, and the fact that spins point in a given direction, is what leads to ferromagnetism. In other words, the way a magnet works is based on spontaneous symmetry breaking!

What we observe in practice is that there is spontaneous magnetization at low temperatures and no magnetization at high temperatures. And there is a *critical* point, somewhere between these two temperature extremes, at which the magnet will undergo a phase transition between these two states (Fig.30).

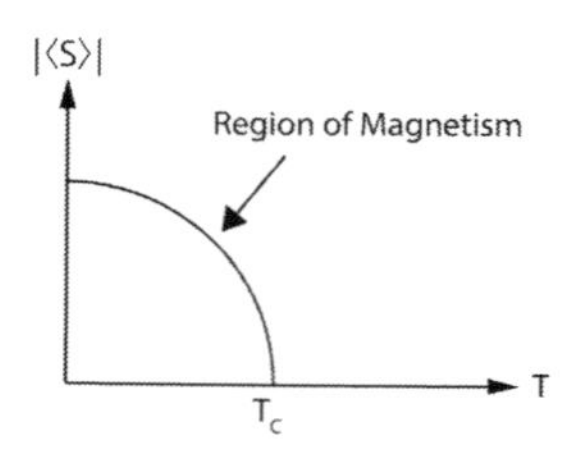

FIGURE 30. The magnitude average spin $|\langle S\rangle|$ is non-zero below a critical temperature T_C.

This might bring to mind the earlier example of the marble

in a bowl with a round hill at the bottom in the very center. The symmetric configuration, with the marble sitting on top of the hill, has a higher energy than the configuration in which the marble is at the bottom of the bowl where the symmetry has been broken. Ferromagnetism, similarly, arises in materials at a lower temperature, after symmetry has been broken. We owe the magic of magnets to spontaneous symmetry breaking!

The Square Puzzle

Puzzle

Four cities are located at the four corners of a square. The distance between adjacent cities is 100 miles. Your task is to come up with a highway system that connects all cities to each other with minimal cost. The cost for building a highway is $100K per mile, so you really want to figure out the minimal total length for the highway. Note that you are not required to have the shortest path between any two cities, and the order in which the cities get connected via the highway system is entirely up to you, just so long as you achieve a minimum total cost. All you need to make sure is that you can get from any city to any other using the highway system. Is there anything peculiar about the solution you find?[14]

Solution

(1) We can show, without too much difficulty, that the minimal network must be a graph with straight lines with cities at some of the vertices. Indeed, since lines are the shortest paths between two points, there is no reason to follow roads that are not straight

(2) Next, let us show that if any vertex of the graph has three edges, then the angles between the three edges ending on the vertex must be 120°. If we consider a triangle near this vertex, where the three vertices of the triangles are on the three edges a unit length away from the vertex, in this case, we can minimize the total length of the highway by finding the minimum road network between the vertices of this *triangle* (and replace this portion of the original graph by this minimal highway). Suppose that indeed the original graph is already the minimum for such a sum of distances to the vertices of the triangle. What does it mean that this vertex minimizes the sum of the distances? It means that when you move the vertex by a small amount, it does not change the total length. Let $\vec{e_i}$ denote three unit vectors connecting the vertex

[14] As we will see later, this problem is related to the Fermat-Torricelli point of a triangle.

to the vertices of the triangle (Fig.31). If we move the vertex by an arbitrary small $\vec{\delta}$, it is not difficult to see that the total change in length is $(\vec{e}_1+\vec{e}_2+\vec{e}_3)\cdot\vec{\delta}$ as $\vec{\delta} \to 0$ (exercise left to the reader). This change in length should be 0 for an optimal choice of highway. Thus, the sum of unit vectors $\vec{e}_1 + \vec{e}_2 + \vec{e}_3 = 0$ since this should be true for any choice of $\vec{\delta}$. This in turn implies that the angles between the unit vectors must be 120° (which can be seen by squaring the relation $\vec{e}_1 + \vec{e}_2 = -\vec{e}_3$ leading to $\vec{e}_1 \cdot \vec{e}_2 = -\frac{1}{2}$).

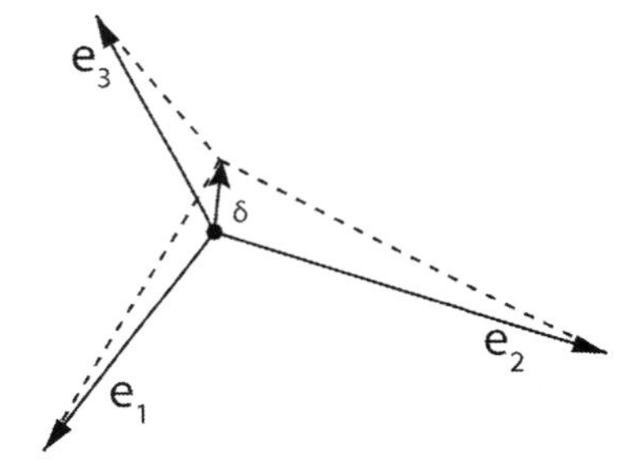

FIGURE 31. For an optimal highway if we move any junction by a small vector $\vec{\delta}$ should lead to no change in total length.

What if a vertex has degree 4, meaning that four edges converge at a single point? Pick two adjacent vertices and the central vertex. We can view this as a *special case* of the previous argument where one point is moved to coincide with a vertex of the triangle. But, as we saw, this can always be improved if the angle is not 120°, but you cannot have four angles of 120° meeting at a vertex. And similarly for higher number of edges.

Using these ideas, it is not hard to see that the only possibilities are those depicted below (Fig.32).

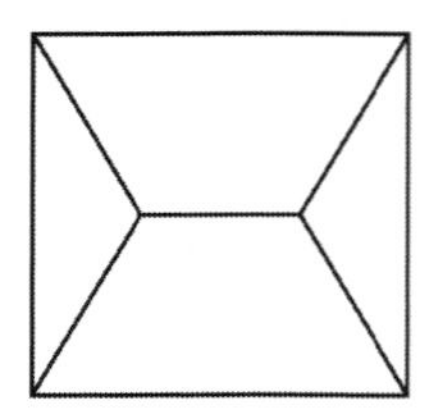

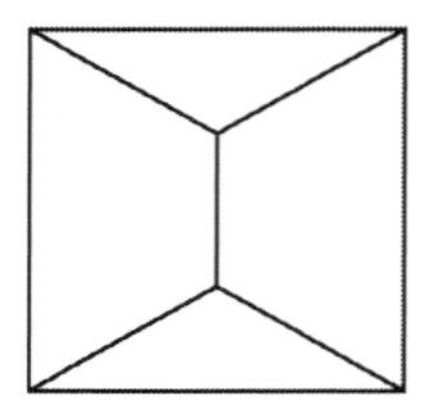

FIGURE 32. There are two optimal highways, neither of which enjoys the full symmetries of the square.

Note that there is spontaneous symmetry breaking because the solutions we find do not enjoy the 90 degree rotation symmetry of the square (even though they enjoy some of the reflection symmetries of the square). Because the symmetry of the square is partially broken in the solution, there is more than one possible solution. You can choose any given solution and transform it by the symmetry that was broken, and thereby get a new solution.

A physical analog of this puzzle is when soap bubbles form, they minimize surface area. A rigid frame dunked in soap water will form soap bubbles at 120° angles.

Alternate Puzzle. We can have the same setup as in the previous puzzle, except that the four cities are on a rectangle. The width of the rectangle is not exactly the same as the height. What would our solution(s) be? What is different about this puzzle from the previous?

Solution

Imagine the width of the rectangle is really long compared to the height. The solution now looks different–a straight line from left to right where at either end it splits to connect to the two cities. As we vary the width until it is almost the same as the height, we still have one unique solution. As the dimensions become exactly a square, we have symmetry breaking, like the previous problem. And as we vary the width until it is smaller than the height, we have a unique solution again, but with an opposite orientation.

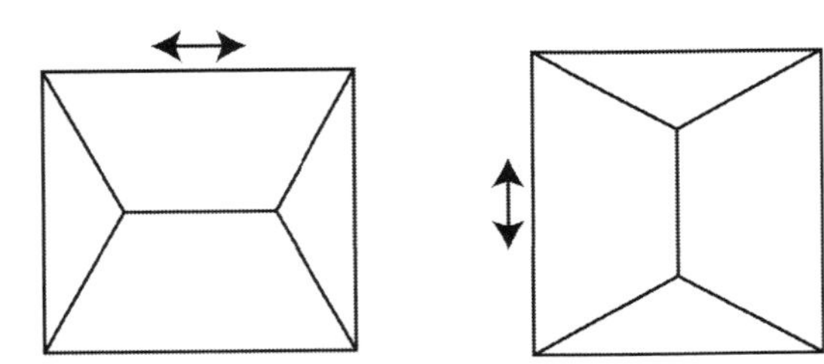

We can clearly see where the two solutions of the previous puzzle come from. This illustrates that the former puzzle has a certain "unstable" quality to it with two solutions switching places as we go from a square to a rectangle by increasing the height or width.

Symmetry Breaking and the Higgs Boson

What is the Higgs particle and what does it have to do with spontaneous symmetry breaking?

We will start off our discussion with a few rather technical equations, geared to the mathematically minded, though others may want to skip over this bit. Those of you who are familiar with the three-dimensional Laplacian might know there is also a four-dimensional Laplacian operator:

$$\Box := \frac{1}{c^2}\frac{\partial^2}{\partial t^2} - \sum_{i=1}^{3} \frac{\partial^2}{\partial x_i^2},$$

where c is the speed of light. The solutions of such an equation are waves. Indeed, think back to the one-variable spatial case.

$$\left(\frac{1}{c^2}\frac{\partial^2}{\partial t^2} - \frac{\partial^2}{\partial x^2}\right)\phi = 0.$$

We can write this as:

$$\left(\frac{1}{c}\frac{\partial}{\partial t} + \frac{\partial}{\partial x}\right)\left(\frac{1}{c}\frac{\partial}{\partial t} - \frac{\partial}{\partial x}\right)\phi = 0.$$

Therefore, we can write this as $\phi(x,t) = f(x+ct) + g(x-ct)$, a sum of a "left-moving" and a "right-moving" wave moving with the speed of light c.

So what does this have to do with the Higgs particle? In the universe's earliest moments, at the start of the Big Bang, every particle was effectively massless, moving at the speed of light. They can also be viewed as waves moving with the speed of light satisfying the 4-dimensional version of Laplacian discussed above. But theory tells us that as the universe cooled during that first tiny fraction of a second, it went through a phase change–not so different from steam condensing into liquid water–and something called the "Higgs field" filled all the space like an invisible ocean. Particles picked up mass through their interactions with this new field, acquiring an additional term in the process: $\alpha_i \phi(x, y, z, t)$ where ϕ is the Higgs field. The wave functions of these now massive particles had to satisfy the equation

$$\left(\Box + (\alpha_i \phi)^2\right) \Phi = 0$$

where α_i depends on the particle, and the $m_i = \alpha_i \phi$ can be considered the mass of the particle.

We can think of ϕ as the analogy of the average spin S, or the position of the ball in the hemisphere bowl. In that case, the most symmetric point–and indeed the only point that would be unchanged by rotation–lies at $\phi = 0$. This symmetry is intact, in other words, only when $\phi = 0$. That situation, in turn, only occurs when the temperature is large, in which case the Higgs field is zero and particles are massless. In other words at high temperature the potential for the Higgs field is like a hemisphere with no dimple at the bottom. However, as the

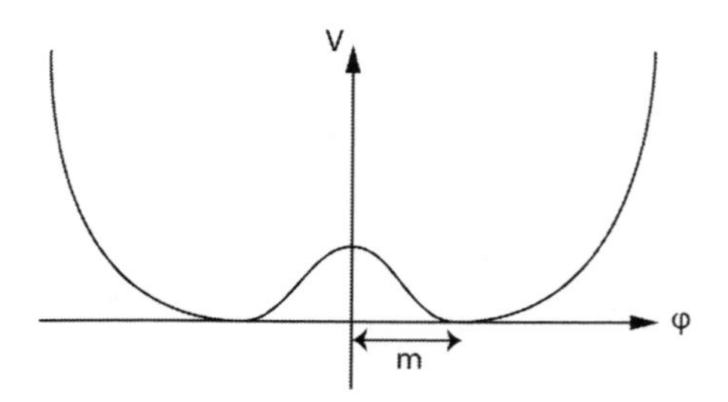

FIGURE 33. The Higgs field has a potential V that looks exactly like the dimple at the bottom of a bowl. The broken symmetry forces the potential minimum away from 0, which leads to mass.

universe cooled down after the Big Bang, it underwent a phase transition: ϕ's energy finds a minimum away from zero, and the $\langle\phi\rangle$ becomes non-zero too, giving rise to non-zero mass (Fig.33). How much mass the particle acquires depends on the strength of its interaction with the Higgs field (captured by α_i) and how far away from the center $\langle\phi\rangle$ lies.

So the universe and the massive particles in it arise from spontaneous symmetry breaking. But how can we prove experimentally that this is the actual mechanism by which particles actually acquire mass?

Think back to the ball in the bowl. If you push the ball away from the bottom and up the hill slightly, it will roll back down and oscillate. Analogously, you can try to push the Higgs field ϕ a bit, and observe the resulting wave. But quantum mechanics teaches us that particles are the same as waves. So if we manage to create a wave associated with the Higgs field, a particle would also be created, which is something we might (and did, in fact) see. This is the Higgs boson, which some members of the news media have dubbed the "God particle."

How should we view this? The Higgs field, as we said, is like an invisible ocean that fills space with some non-vanishing value of ϕ, as the temperature of the universe cooled down after the Big Bang. The interaction of particles with this ocean is what gives them mass and slows them down. How could you show that the ocean really exists if you cannot see it? Well, you could try to "pinch" the ocean to make the Higgs field move up and down the hill. One way to accomplish this would be to collide two particles at a high speed and this leads compressing a tiny patch of this invisible ocean at the point that they collide, which would demonstrate its existence by creating Higgs waves which can be interpreted as a particle. At the Large Hadron Collider at the CERN physics laboratory, high-energy colliders are able to do just that–smash two protons together with enough energy to demonstrate this effect. A Higgs particle can be created in this process, which is exactly what CERN experimentalists saw.

That momentous discovery, announced to the world on July 4, 2012, fulfilled a prediction made nearly 50 years earlier. Theorists' ideas of how particles acquire mass were finally confirmed, and the last piece of the Standard Model of physics–the last expected though still unobserved particle from that theoretical framework–was finally found. This was part of a longstanding learning process in which physicists have come to recognize the power of spontaneous breaking of symmetry. The Earth does move, in spite of the arguments raised two thousand years ago by ancient Greek logicians who had concluded, through symmetry arguments, that it must be stationary. We have now learned how important symmetry *breaking* is. Because of it, the universe is not populated with massless particles that would fly around at the speed of light and could never slow down long to form stars, planets, galaxies, black holes, or any of the wondrous objects we see in the universe, including ourselves.

Grand Unification of Forces

If we are in a situation where symmetry is broken, we may not recognize that there even is a symmetry lurking somewhere in the background. If we lived, for example, at the bottom of a valley with circular symmetry, similar to the shape of the potential for the Higgs field similar to the shape of the ridged bowl (as in Fig.26), we may not appreciate the existence of a circular symmetry. This happens to be the case in our thinking about the forces around us as well. In addition to gravity, there are three other known forces: the electromagnetic, strong and weak forces. Electromagnetic forces are familiar. The strong forces, which bind the quarks to make protons and neutrons, are not so readily discerned. Weak forces–which are responsible for the radioactivity seen, for example, in β decays–are also largely concealed in everyday life. A quark, however, experiences all the known forces. These forces have different strengths, which can be assessed as follows: We fix a small distance, say $10^{-15}cm$, which is equivalent to the energy of $1GeV$ for a photon of that wavelength, and then compute the ratio of the different forces that a quark exerts on another quark, $10^{-15}cm$ away. The ratio of these forces is given by the ratio of the square of the corresponding charges g_i^2. It turns out that in such a scale:

$$g_{strong}^2 >> g_{weak}^2 >> g_{electromagnetic}^2$$

The $g_{electromagnetic} = e$ the familiar electric charge. So these forces certainly appear to be very different, at least in terms of their relative strengths. However, if we continue to ask that question at shorter and shorter distances, the corresponding charges of the three forces change. The charges, moreover, seem to become the same at a distance scale of about $10^{-30}cm$ (corresponding to an energy scale of $10^{16}GeV = M_{GUT}$). See Fig.34.

This is known as the "grand unification" of forces. At higher energies, in other words, the symmetry between forces is restored, whereas at lower energies it appears to be broken. At high energies they seem to unify to one force![15]

[15]Each of the three forces is associated with a group. The strong force is

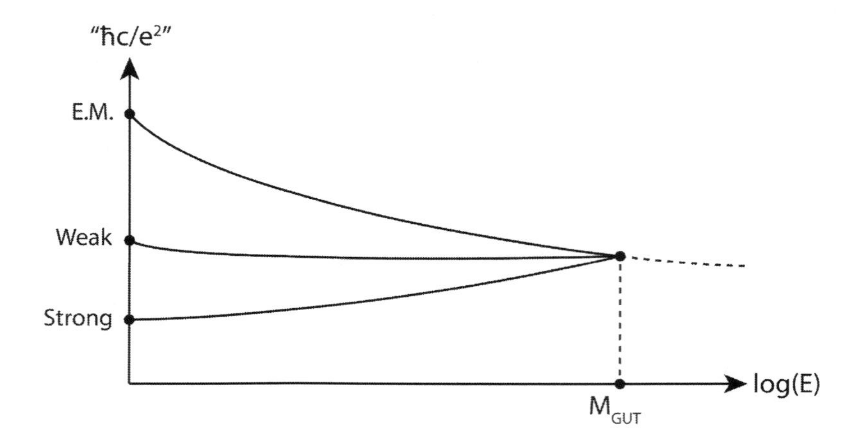

FIGURE 34. Grand unification of forces: At shorter distance scales, which corresponds to higher energies, the charges of the three different forces become the same hinting at the unification of forces.

---

associated with 3×3 matrices known as $SU(3)$ (rotation group in 3 complex dimensions), the weak force with 2×2 matrices known as $SU(2)$ (rotation group in 2 complex dimensions) and the electromagnetic force with a 1×1 matrix known as $U(1)$ (phase multiplication). However, when we shrink the distance enough, the three forces and their respective groups unify into a single group. One popular model, the Georgi-Glashow model, suggests that they unify into a group of 5×5 matrices known as $SU(5)$ (rotation groups in 5 complex dimensions), where the $SU(3)$ and $SU(2)$ come from the 3×3 and 2×2 diagonal blocks in it. The $U(1)$ comes from an overall diagonal, which is orthogonal to the other two blocks.

$$SU(5) \supset SU(3) \times SU(2) \times U(1).$$

Superconductivity

Superconductivity is another example of symmetry breaking. Superconductivity is a property of certain materials, which lose all electrical resistance at sufficiently low temperature, meaning that, once switched on, electric currents would never die. It turns out that the explanation of this phenomenon is based on the same potential as the Higgs potential or the ridged bowl. The analog of the Higgs field in this case is a complex field ρ with potential such that $|\rho| = A$ at the minimum. The current in a circular superconductor comes in discrete units, or quanta, which can be thought of as windings of the phase, of ρ, at the bottom of the circular potential (Fig.35): $\rho = A \cdot exp(i\phi)$, and ϕ winds around the "Mexican hat" potential n times as we go around the circular loop parameterized by the angle θ

$$\phi = n\theta.$$

As it turns out the current $I \propto n$. Therefore the strength of the current is directly proportional to the number of windings of the phase of ρ.

The current, I, is stable for the simple reason that current comes in quantized chunks. Because the current is wrapping around the bottom of a hill, as can be seen from Fig.35, you would have to lift it up in order to unwrap it and that costs

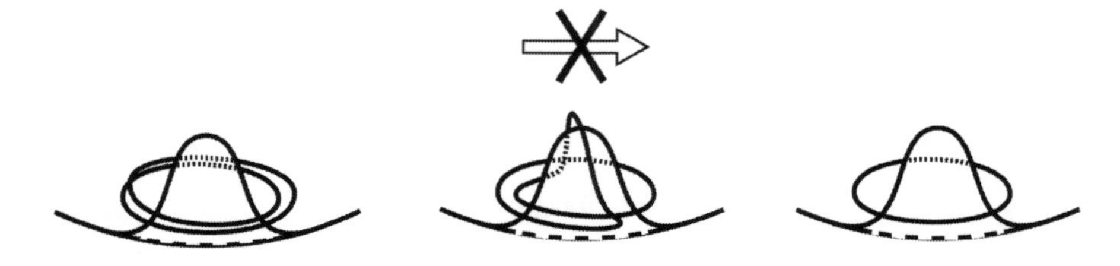

FIGURE 35. Superconductivity can be viewed as arising from the winding of the phase of a field as we go around the circuit's loop at the bottom of its potential. The current is proportional to the winding number of the phase and this is hard to destroy because of the potential barrier. This leads to persistent currents of superconductors.

energy. It takes energy, in other words, to undo the windings and change the current, which is why the current in a superconductor, once set, is inclined to stay where it is.

Rigidity

How about another example? If you push one side of a rigid object, then the other side moves. You are probably not impressed by this observation, which might seem rather commonplace but is actually quite amazing. Somehow you have managed to miraculously *transport* force from one point to another. What is happening in the underlying physics when you do that? The objects are formed of crystals. The fact that they have fixed positions breaks translational symmetry. The breaking of the translation symmetry of the atoms leads to rigidity as a resulting property. The broader point, perhaps, is that many physical phenomena, from the esoteric to things we observe every day, are consequences of symmetry breaking.

Handedness

One of the things we do not understand from the perspective of spontaneous symmetry breaking is the "handedness" of the universe. Some particles have a sense of "handedness" associated with the orientation of their spin relative to their motion. This breaks the parity symmetry (reflection off a mirror). However, this looks like the breaking of symmetry "by hand," meaning that the particles never had it in the first place. A more elegant solution would be for the symmetry to be in place originally and then disrupted through a natural process. We hope that in a complete theory, such as string theory, handedness will be spontaneously broken, as opposed to the more contrived situation in which symmetry is artificially broken by hand.

This is somewhat reminiscent of the puzzle involving four cities at the corners of a square or rectangle, where the highways have already been built. The asymmetry in that problem is due to the constraints on the budget, which corresponds in

FIGURE 36. In this illustration of the geometry of knives and glasses, which arrangements of knives over the glasses can lead to supporting a heavy bottle?

the physical universe to the constraint of low energy.

Puzzle [16]

You are given three knives of length L and three glasses that are equidistant from each other by length (slightly larger than) L (Fig.36). Arrange the knives in some configuration to support a heavy bottle on the glasses.

Solution

One must break the symmetry to solve the problem. We can rest the bottle over the triangular region formed by the knives as in Fig.37.

[16]This puzzle was shared by Brian Greene.

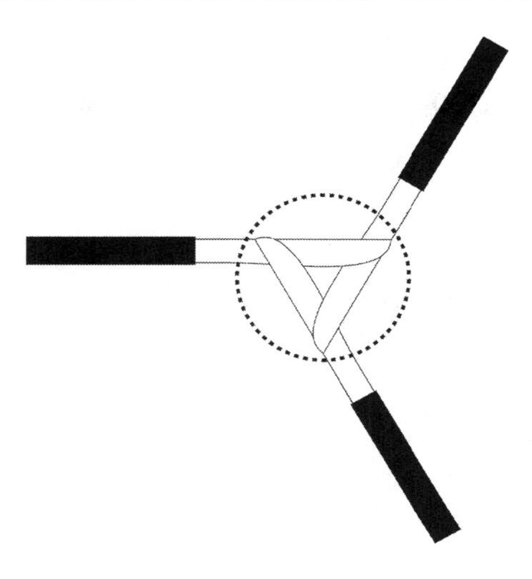

FIGURE 37. The geometry of knives (the three glasses not drawn) supporting a bottle breaks the symmetry and introduces handedness.

This imparts a handedness to the situation, breaking the original symmetry that once existed between the three glasses.

Puzzle

Suppose we have a circular pond of radius R. There is a duck sitting at the center of the pond. A fox, which cannot swim, sits on the edge of the pond, and, naturally enough, it wants to eat the duck. The duck needs a strategy to get to land without being eaten by the fox, where it can then fly away and avoid being eaten. The fox moves x times faster than the duck, where $x > 1$. Can the duck escape? If so, what strategy should the duck adopt?

Solution

Let r_1 be the radius where the duck can maintain a higher angular velocity than the fox. Thus $r_1 < R/x$. This means that when the duck is in the circle of radius r_1, it can move until the fox is on the opposing side of the pond. Let r_2 be the radius at which the duck can make a straight run for the coast and escape, assuming the fox is on the opposing side of the pond. The duck needs to move $R - r_2$ distance, whereas the fox needs to move πR distance. Thus we need $R - r_2 < \pi R/x \Rightarrow r_2 > R - \pi R/x$. If the regions of r_1 and r_2 intersect, then the duck can escape. If not the fox will always be able to eat the duck. The condition for this transition is $R/x = R - \pi R/x \Rightarrow x = \pi + 1$. So the duck can escape if $x < \pi + 1$.

This is an interesting problem because we start with a symmetric situation, but the duck is forced to break circular symmetry by picking a direction to go around the circle to get to the opposite side of the pond. When $x > \pi + 1$, no matter how the duck breaks circular symmetry, the fox will still eat it. So we only have a narrow range of values of x, namely $1 < x < \pi + 1$, where circular symmetry is broken in a way that would permit the duck's survival!

4. **The Power of Simple and Abstract Mathematics**

Laws vs. Constraints

When it comes to solving physics problems, there are two kinds of inputs that are typically drawn upon: First, there are the constraints placed on a problem sometimes referred to as boundary conditions. This includes factors imposed upon the situation by the environment that might not seem, at first glance, to be very deep. Consider a ball, for instance, accelerating down an inclined plane. Without knowing anything about physics, we can say that the ball is going to be somewhere on the inclined plane. These are aspects of the physical phenomena dictated by the environment which can be viewed as constraints. Constraints loom rather large in classical mechanics when one wishes to describe the motion. This is also known as "kinematics." Second, there are physical laws, such as those formulated by Newton or Einstein, which seem to be much more fundamental. "Dynamics" addresses forces influencing the motions of objects and systems as a whole, where physical laws play a bigger role.

Part of the discussion in this section will focus on what might appear to be the duller side of physics, constraints, but as will hopefully become clear in the course of this chapter that is not necessarily the case. Some of the ideas we are going to take up here can manifest themselves in very deep ways. And we may find, at a very basic level, that the distinction between laws and constraints disappears and that many of the things we attributed to principles could actually emerge from constraints.

Mathematically, we will view topology as an analog of these general physical constraints. Topology describes the global, qualitative aspects of a space, its general features, as opposed to geometry, which delves into the details of a space, involving distances, precise shapes and so forth. Continuity, which is a basic idea in the context of topology, has a natural link to the fact that the laws of physics are continuous: If you change things a bit, the outcome will typically not change drastically.[17]

Puzzle

There are 117 players in a round-robin, single-elimination tournament. How do we set up the tournament to arrive at a winner with a minimum number of total games? What setup, conversely, would lead to the maximum number of games?

Solution

You could fall into a trap by trying to spell out the details of the tournament; that would be a complicated task and wholly unnecessary. The answer turns out to be quite straightforward: there are *always* 116 games. The reason for that is simple: each game eliminates one player, and 116 players must be eliminated.

[17]Though there are exceptional cases such as chaotic physical systems where this is not strictly true.

In other words, 116 games must be played. That is both the minimum and maximum number. Any question about how to structure the tournament is a smokescreen. So do not be fooled; constraints drive this problem. It is unnecessary to try to figure out the optimal way of pairing up the players in the tournament, because the answer can be found, more simply, in the puzzle's original setup.

Puzzle

64 teams are in a double-elimination tournament. How many games will the tournament consist of?

Solution

In each game there is exactly one loser. Each team needs to lose twice to be eliminated. Therefore, there are 63 teams that lose twice, and the champion can either lose once or no times at all. There can be either $2 \cdot 63 = 126$ or $2 \cdot 63 + 1 = 127$ games in the tournament. The original framing of this problem, again, imposes enough restrictions to force a limited set of possible answers.

Puzzle

A chocolate bar is arranged in a (continuous) 5×20 grid (Fig.38). Two players take turns cutting the bar along either

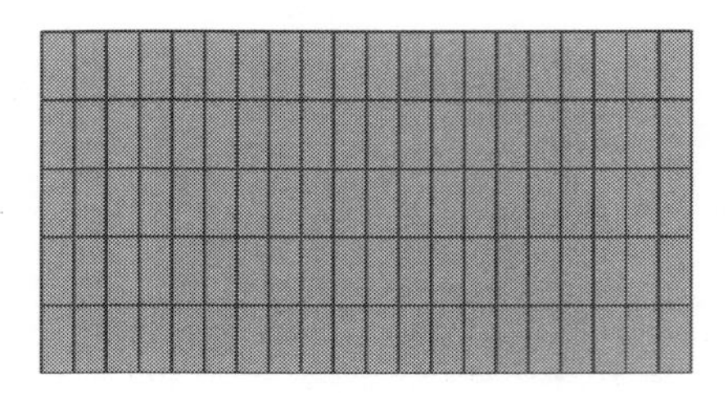

FIGURE 38. What strategy leads to winning the chocolate bar game? The game involves two players taking turns cutting the chocolate bar along the edges and the last one to make the last possible cut gets all the chocolate!

a vertical or horizontal line. (Multiple cuts–made by stacking chocolate pieces on top of each other or by continuing to sweep the knife across more than one piece–are not allowed.) The last person able to make a cut gets all of the chocolate. What is the winning strategy?

Solution

The first player always wins. At the beginning there is just one piece; at the end there are 100. Each break increases the number of pieces by exactly one. The number of pieces goes from 1 to 100, so there will be 99 cuts in total. That is an odd number, meaning that the player who goes first ultimately gets to make the 99th cut. Again, it is difficult and unnecessary to try to map out the entire course of the game. The original conditions (or constraints), along with logic, can guide us towards a solution.

A Primer on Complex Numbers

Before getting to the next round of puzzles, we require some math background, which we will try to build up now.

The complex numbers, $\mathbb{C}$, can be represented as points in the plane. In polar coordinates, we may write any $z \in \mathbb{C}$ as $z = re^{i\theta} = r(cos(\theta) + isin(\theta))$ where r is the distance to the origin and θ is the polar angle. The complex conjugate $z^* = re^{-i\theta}$. Given two complex numbers $z_1 = r_1 e^{i\theta_1}$ and $z_2 = r_2 e^{i\theta_2}$, their product is

$$z_1 z_2 = r_1 r_2 e^{i(\theta_1+\theta_2)}.$$

Note that $zz^* = r^2$. We sometimes denote $|z| = r$.

We next prove a theorem about complex numbers. The reason we present this theorem here is not only that it illustrates the power of topological arguments in mathematics, but in addition this example mirrors what we will find later in this chapter in physical examples.

Fundamental Theorem of Algebra

Let $f(z) = z^n + a_{n-1}z^{n_1} + \ldots + a_0 = 0$. If $n \geq 1$, then then $f(z)$ has a solution in $\mathbb{C}$.

A simple consequence of this is that a polynomial of degree n with coefficients in $\mathbb{C}$ has exactly n solutions, counted with multiplicity. The fundamental theorem lies at the root of the ubiquity of complex numbers. So how do we go about proving this theorem? Suppose $f(z)$ had no solutions. We now show that this assertion leads to a contradiction. If $f(z)$ had no zeroes, the function

$$g(z) = \frac{f(z)}{|f(z)|}$$

exists where $|f(z)| = \sqrt{f(z)f(z)^*}$ and $f(z)^*$ is the complex conjugate of $f(z)$. Note that this division is allowed precisely because we have assumed that $f(z)$ is never 0.

The function $g(z)$ was constructed to have the property that $|g(z)| = 1$ for all $z \in \mathbb{C}$. In other words, $g(z)$ lies on the unit circle. Therefore, as we vary the value of $z \in \mathbb{C}$, $g(z)$ maps the whole complex plane to the unit circle.

Consider the image under g of an extremely large circle, whose radius is much, much bigger than all of the coefficients of the polynomial. Then, to a very good approximation, $f(z) \approx z^n$ for $|z| \gg 0$ because the other powers are much smaller by comparison. Similarly, $g(z) \simeq \frac{z^n}{|z|^n} = e^{in\theta}$ for $|z| \gg 0$. The conclusion is that for really big circles, g wraps the big circle on the complex plane around the unit circle n times (Fig.39), because as θ varies from 0 to 2π, g varies from 1 to $\exp(2\pi in)$.

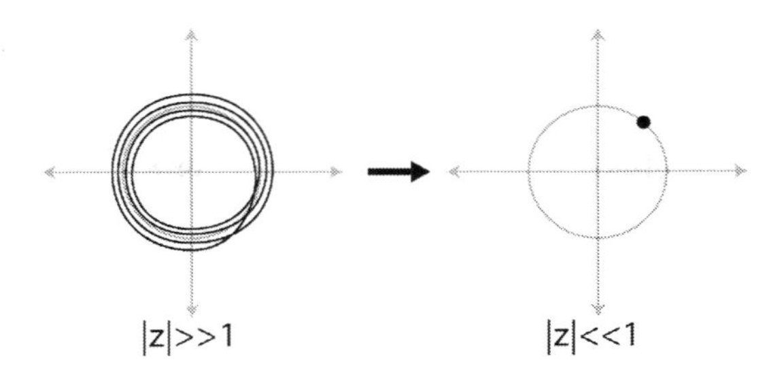

FIGURE 39. Fundamental theorem of algebra can be proven using a simple topological argument based on conservation of winding number

Note, however, that g is continuous. Since it winds around n times for really large circles, and g varies continuously of the complex plane, this number cannot jump. Therefore, it must wind around n times even as we gradually shrink the circle, and even until we get to small circles as well. But as we contract, z gets very close to 0. So $f(z) \approx a_0 = g(0)$. At the end, in other words for $|z| << 1$, $g(z)$ has no winding and it corresponds to a single point $g(0)$ on the circle, which is a contradiction for the winding number to be $n \neq 0$. Therefore, the assumption that $f(z)$ does not have a zero, which led to the construction of the continuous function $g(z)$, cannot be true.

In reality, we know that f has n solutions, which represent discontinuities of g. The winding number jumps by one at each of these discontinuous points (where the denominator of g is 0), which provides another explanation for why there are n solutions. Another way of saying this is that if a is a zero of f we can divide f by $(z - a)$ leading to a polynomial of one lower degree and then we can repeat the above argument. This leads inductively to the conclusion that f has n zeroes (counting the zeroes with multiplicity if they are repeated).

Puzzle

Consider the temperature T along the equator of the Earth. Assume that T is a continuous function of the position along the equator. Show that at any given time there are at least two diagonally opposite points along the equator which have exactly the same temperature. (Hint: You do not need to know any facts about thermodynamics. Or about meteorology or geography).

Solution

Define the function $\tilde{T}(\theta) = T(\theta) - T(\theta + \pi)$, which is the difference between the temperature at a point and the diametrically opposite point. If $\tilde{T}(\theta) = 0$, for any θ, we have found the desired point. Otherwise, observe that $\tilde{T}(\theta + \pi) = -\tilde{T}(\theta)$, so if $\tilde{T}(\theta_0) \neq 0$ for some θ_0 (say it is positive), then it has opposite sign when you go to $\theta_0 + \pi$ (negative). Therefore, $\tilde{T}(\theta)$ must be zero (see Fig.40) at some point between θ_0 and $\theta_0 + \pi$ due to continuity (also known as "the intermediate value theorem"), because $\tilde{T}(\theta_0)$ and $\tilde{T}(\theta_0 + \pi)$ have opposite signs.

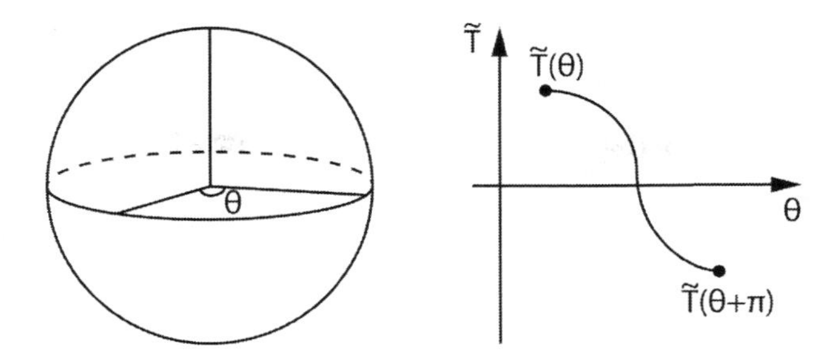

FIGURE 40. Temperature difference $\tilde{T}$ at a point and the diametrically opposite point will change sign as we go to the diametrically opposite point because the difference switches sign. The temperature of the two opposite points are the same where this function crosses 0.

This statement seems amazing. But as we see from the foregoing argument, it follows from some rather trivial topological/continuity considerations.

Puzzle

A monk climbs from the base to the top of a mountain from 8 a.m. to 8 p.m. On the next day, he climbs down from 9 a.m. to 7 p.m.. Show that there is a moment at which he is in exactly the same point at exactly the same time of day.

Solution

If you plot the graphs (Fig.41) of distance versus time for his trip up and down, you'll see that they must cross somewhere.

(If the graphs do not intersect anywhere, the monk has still not made his way down the mountain, in which case we should send out the search parties immediately!) A more physical way of seeing this is that you can imagine that another monk retraces exactly the path the monk took the previous day, and clearly the two monks will cross paths as they pass each other, where one is descending and the other is ascending.

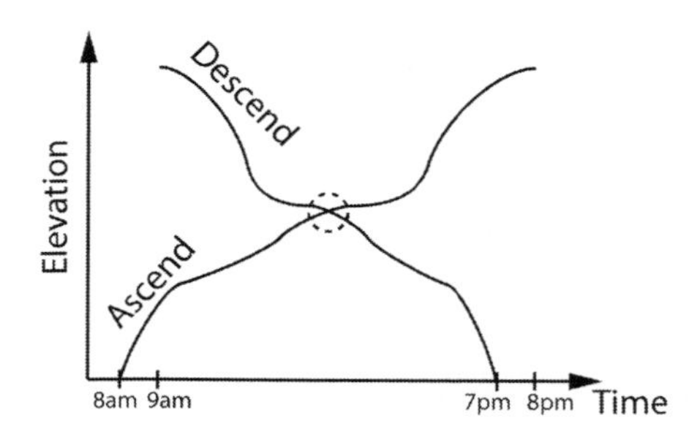

FIGURE 41. Monk will be at the same elevation as in the previous day at some point, as can be seen by continuity of the elevation as a function of the time of the day.

Puzzle

Building on the previous puzzle about temperature: Is there a point on Earth that has the same temperature *and* air pressure as its antipodal point right now?

Solution

The answer, again, is yes. The argument is a bit more technical but it is again based on the principle of continuity and winding numbers not changing. Consider the vector-valued function $\vec{f}(x) = (P(x), T(x))$, where P is the pressure, T is the temperature, and x denotes a point on Earth. Then consider the function $\vec{g}(x) = \vec{f}(x) - \vec{f}(-x)$: A zero of this function corresponds exactly to the point where pressure and temperature are identical to its antipodal point. Let us assume this never happens and try to arrive at a contradiction. If $\vec{g}(x)$ never vanishes, then we can divide by its norm and consider instead the normalized vector

$$\vec{g}(x) = \frac{\vec{f}(x) - \vec{f}(-x)}{|\vec{f}(x) - \vec{f}(-x)|}.$$

This maps the sphere to the unit circle because the vector has unit norm. Consider a foliation of circles on the sphere along longitudinal lines (Fig.42). Since $\vec{g}$ is continuous, the image of circles shrinking to the north pole shrinks to a point, so the winding number is 0. Therefore, the winding numbers are all 0 by the same continuity argument as before, including the winding number of the circle corresponding to the equator.

But let us consider where $\vec{g}$ maps the equator to. The image of the half-circle between antipodal points A and B is some arc with winding number of the form $n + \frac{1}{2}$; the half comes from the fact that $\vec{g}(x) = -\vec{g}(-x)$, so points A and B must end up on

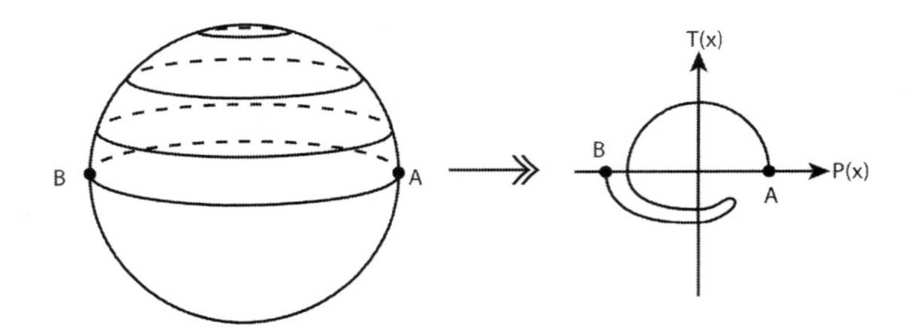

FIGURE 42. As we move from point A to point B on the equator, $\vec{g}(x)$ moves from one side of the unit circle to the other. The winding number of $\vec{g}(x)$ as it moves from A to B is $n + \frac{1}{2}$.

different sides of the unit circle and as we continue going around the equator till we get back to A we get exactly the negative of the first half. Therefore, the image of the whole equator has winding number $2(n + \frac{1}{2}) = 2n + 1$ and since it is an odd number it cannot be 0. Thus we arrive at a contradiction because we had expected it to be 0 based on continuity. This is exactly the same reasoning we applied in our proof of the Fundamental Theorem of Algebra: The division by $|g|$ must have not been allowed, and g must have a zero, and this is precisely what we wished to show. The constraints in this case, stemming from the geometry of the sphere and continuity, are sufficient to lead us to the answer.

Puzzle

Given a closed curve and any point on the curve, is it possible

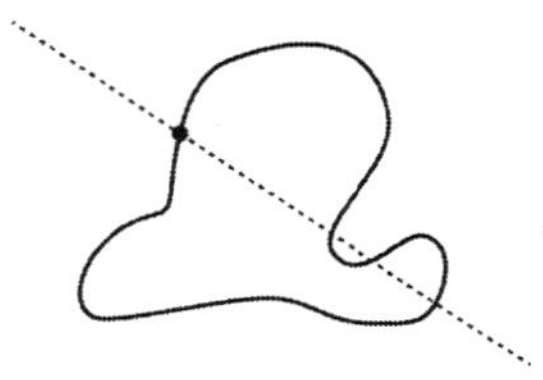

FIGURE 43. Can you always split the area in half by drawing a line through the curve passing through a given point on the curve?

to draw a line through that point that splits the area in half (Fig.43)?

Solution

Yes, given any point on the curve, take the difference between the area on the left-hand side of the line and the right-hand sign of the line as a function of the angle the line makes with the curve. After 180°, the function changes sign as left becomes right and right becomes left. So if the function is positive on one side of the 180° line and negative on the other side, we know by continuity that there must be some point in between where the function is zero. And that is exactly where the line splits the area in half.

Puzzle

Given two closed curves, is it possible to draw a line that simultaneously cuts both into equal areas (Fig.44)?

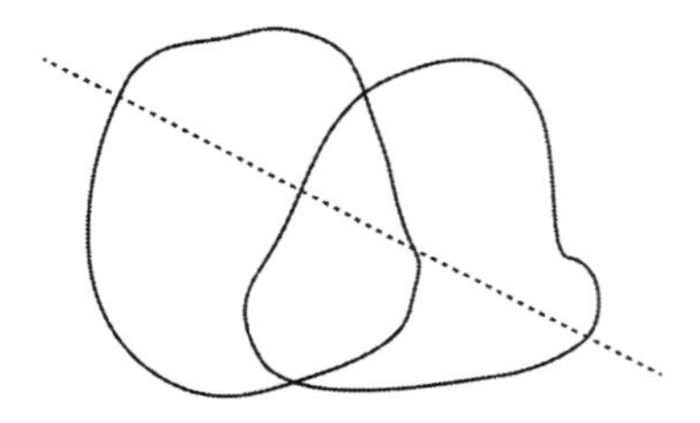

FIGURE 44. Can you always split the area of each of two curves in half by drawing a line passing through them?

Solution

Yes! Pick a point on one of the curves and constrain the line to evenly divide the enclosed area of the first curve. (We know this is possible from the previous puzzle.) Consider moving the point on the first curve, while always picking the line that evenly splits the first curve. We will now consider a function that takes the difference of the area on the second curve that such a line creates as we move along the first curve. When we reach the antipodal point on the first curve–the point that is exactly opposite (and 180°) from the one we started with–the function will once again change signs, going from positive to negative, or negative to positive. Owing to continuity, there must be a point in between where the function is zero–a point where the areas within both curves are simultaneously cut in half.

Puzzle

This puzzle uses the fact that one can have an arbitrarily long gap in integers where there is no prime number. Recall that a prime number is a number that upon dividing by any integer (except 1) leads to a remainder. To convince yourself of this fact, note that $k!+2, k!+3, ..., k!+k$ gives you $k-1$ consecutive integers that are not prime ($k!+n$ divides n when $n \leq k$ because both $k!$ and n divide n).

Show that you can find an integer, N, such than between N and $N+1000$ there are exactly 13 primes.

Solution

Our argument relies upon the simple idea of discrete continuity, which we will now explain. Let $p(N)$ indicate the number of primes between N and $N+1000$. Note that between 1 and 1001 there are more than 13 primes. Therefore, $p(1) > 13$. Note that for any N, $p(N+1)$ differs from $p(N)$ by at most one unit. In this sense, $p(N)$ has a discrete continuity. Since we know there is a large enough M such that $p(M) = 0$ (by the fact that we can have arbitrarily large gaps in primes), $p(N)$, therefore, goes from a number bigger than 13 to 0 as we reach M. By discrete continuity it follows that $p(N) = 13$ for some $1 < N < M$, which is what we wished to show.

Gravitational Lenses

Einstein offered a *geometric* explanation of gravity. Rather than viewing gravity as an attractive force between massive objects, which is how Newton described it, Einstein's theory is based on the notion of curvature. The presence of mass, he teaches us, literally causes the fabric of space and time to curve or warp. The warping of space and time affects the motions of objects nearby, and that, in essence, is the phenomenon we call gravity.

To explore this idea, consider two points on the sphere. There is a unique *shortest* path (or *geodesic*) on the sphere between these two points (Fig.45). Although this path is not "straight" in the conventional sense, it is still the shortest path that lies in the sphere. Actually, there are exceptions: there are infinitely many shortest-length paths between antipodal points. Instead we can start from a given point on the sphere and go in a given direction in as straight a manner as we can. This is what is called a geodesic. The path we thus get will be a great circle on the sphere.

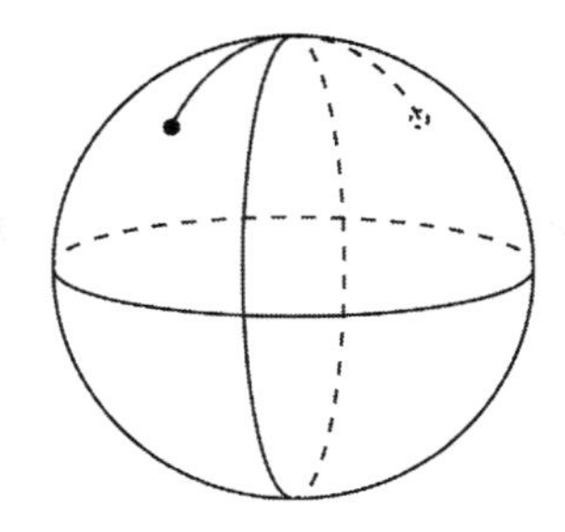

FIGURE 45. There is normally only one shortest path between any two points on the sphere.

Other kinds of situations can arise. On the torus, a donut shaped object as in Fig.46, there are topologically inequivalent geodesics paths between points lying on opposite sides of a cross-sectional circle.

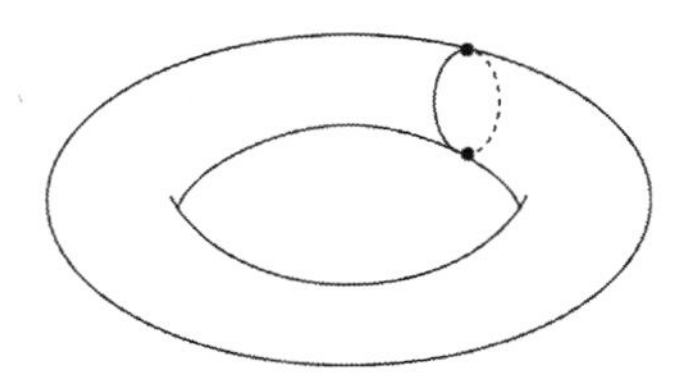

FIGURE 46. On a torus, there are two shortest paths between points sitting on opposite sides of a cross-sectional circle.

Einstein's theory predicts that light always travels in the shortest path between two points. However, the curvature of space-time means that the path may not look straight in Euclidean space. The theory tells us, for example, that the sun, being a massive body, should bend passing light. This prediction was the first experimentally confirmed prediction of Einstein's general theory of relativity, when the bending of light around the sun was indeed observed during an eclipse.

Now this raises an interesting question: can a physical situation arise in which there are *multiple* geodesics giving rise to multiple images of a single object?

Even though he knew that the production of multiple images was theoretically possible, Einstein did not believe it was likely to be observed. But the first example of gravitational lensing was observed nearly 40 years ago, in 1979, when two images of the same quasar (known as the Twin Quasar or Double Quasar)–lensed by a galaxy lying between the quasar and Earth–were spotted by astronomers at an Arizona telescope. And countless examples have been detected since then. We will see in the next section that gravitational lensing should, in principle, yield an odd number of images. However, in cases in which some of the light rays are blocked, astronomers on Earth may see an even number of images, which happened with observations of the aforementioned Twin Quasar.

Generically, assuming that no light rays get blocked, there is always an odd number of images. If this number is $2n + 1$*, then exactly* n *of the images are inverted (meaning the orientation is reversed).*

To establish the above point, it might seem that we would have to understand some deep facts about Einstein's theory of relativity–a theory built around a set of rather imposing nonlinear partial differential equations. However, as we shall see, the main thing we need to know is that Einstein's theory respects continuity. Before proving this, we must first establish a bit of math background–for those who are up to some heavy lifting in this area. Let $f : X \to Y$ be a map of spaces. (We will restrict ourselves here to smooth maps between smooth manifolds of the same dimension). This brings up the notion of *degree*, which plays a central role in the following argument.

The loose definition of degree when mapping from X to Y is the number of points (or pre-images) of X that get mapped to each point (or image) of Y. Restating that in mathematical terms, we could define the degree as follows: for some $y \in Y$, the degree is $\#\{f^{-1}(y)\}$.

Example: The map $f : S^1 \to S^1$ given by $\theta \mapsto n\theta$ has degree n.

But complications can arise, as in the following example: Consider a map of two concentric circles, $S^1 \to S^1$, mapping the outer circle to the inner one, except that the outer (closed) curve is not exactly a circle but instead has a small fold in it (as shown in Fig.47). The radial line defines the map from the outer circle to the inner one. At places where the radial lines intersect the folds, the map from the outer to inner circles appears to be $3 - to - 1$ rather than $1 - to - 1$. But the map can still be called degree 1, generally speaking, if we take take into account the fact that the pre-images have different "orientations," which have different signs (+ or -) associated with them, as can again be seen in Fig.47 (the middle point in the figure below will have a minus attached to it). Two pre-images of opposite sign cancel each other out, leaving us with a map that is degree 1.

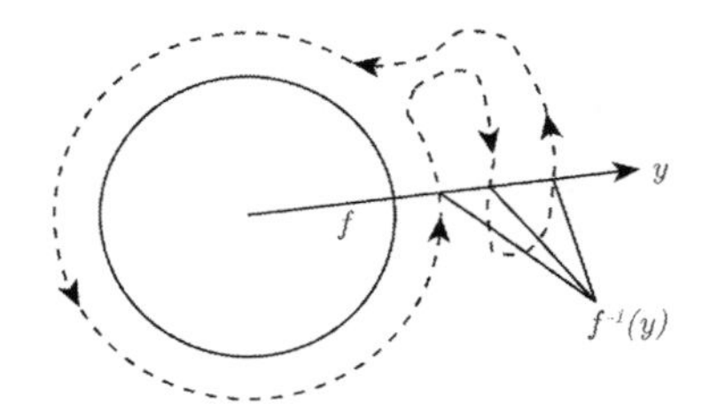

FIGURE 47. The map is defined from the dashed circle to the solid circle by following along radial lines. A map of degree 1 can have a non-standard form that "doubles back" on itself for a bit, thus giving it many pre-images. To account for this, we must account for the orientation of the pre-image as well.

An additional problem remains: there are some points at which there are even numbers of pre-images when two of the opposite sign pre-images coalesce and are about to cancel one another. This occurs for a discrete set, so we can avoid them by a small perturbation of the point we consider.[18]

We will now return to the problem of light rays. We assume that there is nothing to block any light rays from getting to us. We will reformulate the problem so that it becomes, more simply, a question of counting the degree of a map.

[18]The general definition of degree is this: If $f : X \to Y$ is a map of smooth manifolds, then we want to assign ± 1 to the multiplicities. Locally around the points, the spaces look like Euclidean space, and f induces a map, $\mathbb{R}^n \to \mathbb{R}^n$. The sign we attach to each such point is the sign of the determinant of the Jacobian matrix.

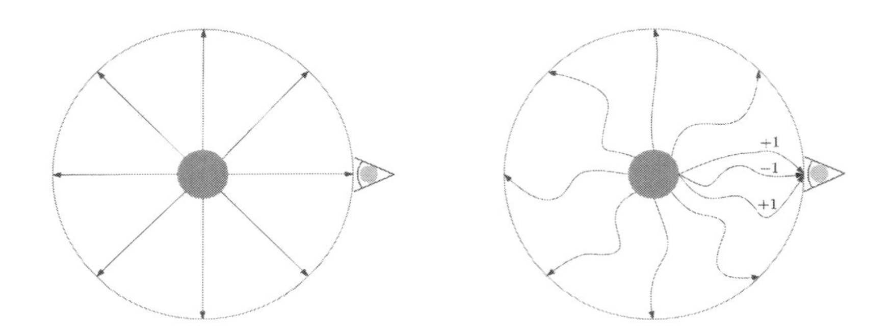

FIGURE 48. The degree of the map that governs the path of light from the star to an external observer is always one. On the left with no other matter in the middle the degree is clearly one. After adding matter in between the star and the observer the map changes, but due to continuity of physics, the degree cannot jump. Therefore the net number of images (taking into account orientation) is still one.

Consider a star whose image we are observing. Consider a big sphere, centered at the star, which passes through us. Now consider another much smaller sphere, again centered at the star but just large enough to include the star itself, like the surface of the star (Fig.48). Consider the map from the small sphere to the big one obtained by tracing out the path of the light ray. This map exists because we are assuming that no rays get blocked so every light ray gets to infinity, thus passing through the big sphere, and it is continuous because the laws of physics are continuous. We are especially interested in finding out the degree of this map. Imagine slowly "turning off" all of the masses in the universe between the star and us. Then the light rays all become straight lines, and the map becomes the identity map $S^2 \to S^2$, and hence has degree 1 (as in the left figure of Fig.48). Now imagine continuously cranking up the masses. The degree will always be one, because the association of degree to a map is continuous due to the fact that the laws of physics are continuous as we change parameters. Since the final degree is one, it means that there must be an odd number of pre-images. That is, we will observe an odd number of images such that when we count them with $+/-$ signs taking orientation, we should get 1. This means that the number of points is $2n + 1$, with n points of negative orientation. Notice that we did not use anything but continuity from Einstein's theory of general relativity!

One point that emerges from this exercise is that many seemingly difficult physics problems can be solved without invoking (much, if any) physics. We have to be careful to find out whether statements in physics are fixed by topology, which we are regarding here as constraints, or are instead based on details of the physical laws.

5. Counter-Intuitive Mathematics

Preliminaries

For better or worse, we are creatures of habit. The experiences we have had make an imprint on us, coloring our perceptions. Sometimes wisdom emerges from those experiences, but we may also pick up some wrongheaded notions along the way. When it comes to mathematics, we may approach a given problem with preconceived ideas as to what the correct answer should be. Although intuition can be valuable at times, it can also mislead us. Simple mathematical thinking, however, can often clarify things. The following anecdote shows what can happen when our intuition leads us astray.

A Joke. –A mathematician, a physicist, and an engineer are trying to prove that all odd numbers are prime.

The mathematician says, "3 is odd, 3 is prime. 5 is odd, 5 is prime. 7 is odd, 7 is prime. By induction, all odd numbers are prime."

The physicist says, "3 is odd, 3 is prime. 5 is odd, 5 is prime. 7 is odd, 7 is prime. 9 is odd, 9 is not prime. An experimental error. 11 is odd, 11 is prime. 13 is odd, 13 is prime, etc."

The engineer says "3 is odd, 3 is prime. 5 is odd, 5 is prime. 7 is odd, 7 is prime. 9 is odd, which $+/-10$ is also prime, etc."

Puzzle

Imagine there is a giant belt wrapped tightly around the equator of the Earth. We open the belt and add one meter to its length. How far above the Earth's surface does the belt rise? Can you pass a piece of paper underneath it? How about a mouse? Or a skyscraper?

Solution

The most naive guess one could make, without relying on any mathematics, is that the belt will rise only a tiny bit, so you could not even pass a piece of paper underneath it. This expectation, which our intuition might lead us towards, turns out to be false. If you imagine lifting the belt uniformly above the Earth so that it again forms a circle, the new circumference will be $2\pi R + 1$, where R is the radius of the Earth. The radius of the new circle that the belt forms is $(2\pi R + 1)/2\pi = R + \frac{1}{2\pi}$, which turns out to be about 16cm longer than R. While 16 cm is not huge, it still seems surprisingly large for this problem, in view of what most people's expectations would likely be. So you could, indeed, pass a mouse under the belt, as well as some larger rodents and even a cat!

An even bigger surprise comes when we consider the situation in which the belt does not have to rise equally in all directions. What is the farthest it could extend? Probably the most naive expectation is that you could pull the belt up at one point as far as it would go and that this is going to lead to a height of about half a meter (folding it in half the extra slack affords an extra half meter in height). It turns out, however, that pulling it up at one point leads to far larger rise above the equator. To actually obtain the amount of rise needs a bit of calculus, which we will now discuss (a reader unfamiliar with calculus may wish to skip it). Let us suppose the belt is not tangent to the surface of the Earth only over an angular size of 2θ (Fig.49). As can be seen from the figure, the extra one meter of belt (as compared to the Earth's circumference) is $\epsilon = 2R\tan\theta - 2R\theta$ and the height of the rise above the equator is given by $h = R \cdot Sec\theta - R$. Taking θ to be small and expanding the functions in power series, we can eliminate θ and find a relation between h and ϵ: $h = \frac{1}{2}R^{1/3}\left(\frac{3\epsilon}{2}\right)^{2/3}$. Note that $dh/d\epsilon \propto \epsilon^{-1/3} \to \infty$ as $\epsilon \to 0$. In other words, the ratio of the height we gain to the length ϵ we add blows up as the length we add goes to zero. For $R =$ the radius of Earth and $\epsilon = 1$, we have $h = 121m$, which is truly counter-intuitive. So we could indeed pass a skyscraper underneath it! The "Big Ben" clocktower would easily fit beneath the belt too–as would the Statue of Liberty, including the base and the tip of torch.

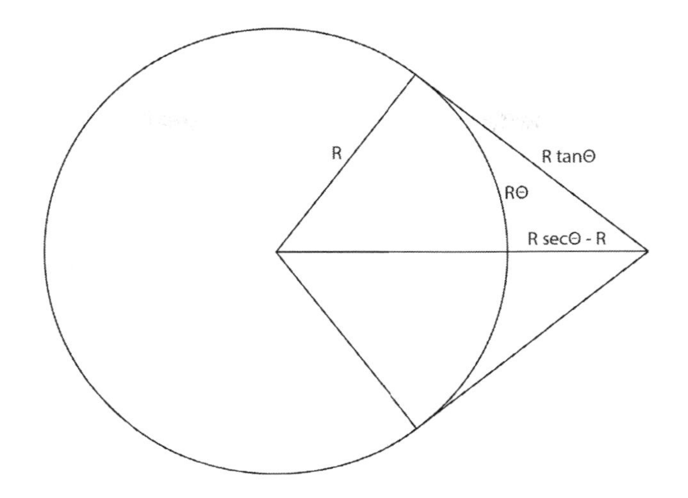

FIGURE 49. The belt pulled up from one side leads to a 121 meter elevation above the equator, even though the length that was added to the rope was only 1 meter.

We can make this more intuitive by the following reasoning: First note that a circle has the maximum area of any shape with a given circumference. So if we deform the circle a small amount to a different shape, its area should not change significantly (because we are already at the maximum of the area). What we have done here in this second solution to the problem is to take all the area between the belt and Earth from the first solution and place it under a narrow corner of the belt. That is why we were able to achieve such a large elevation.

Puzzle

Pick n generic points on a circle and connect all possible pairs of points by straight lines. This divides the interior of the circle into many regions. The question is to find out how many regions that area is divided into for a given number of points on the circumference. For $n = 2$, for example, we get 2 regions, and for $n = 3$ we get 4 regions. Is there a general formula and, if so, what is it?

Solution

For small values of n, here are the results

n	Regions
2	2
3	4
4	8
5	16

This suggests an answer of 2^{n-1}. One could even provide a quick explanation for why that might be true: each new point creates additional lines that cut every region in two, so the number of regions would naturally double each time you add a point.

There's just one problem with that reasoning: the answer it gives is *wrong*. For $n = 6$, there are 31 regions, not 32. Similarly for $n = 7$ we get 57 regions rather than 64. The general answer, expressed in terms of binomial coefficients, turns out to be[19]:

$$1 + \binom{n}{2} + \binom{n}{4}$$

Let us think about the three terms. If there are no lines, there is 1 region to start out. This explains the first term. For each additional line, we get a new region, which explains the $\binom{n}{2}$ because there are that many lines when we have to choose a pair of points out of n points. However, for every group of four points, there is an extra intersection point that adds one extra region This explains the $\binom{n}{4}$. You can convince yourself that there are no other sources of pathologies.

We were tricked initially to believe the answer is 2^{n-1}, because it is so simple and it works up to $n = 5$. Moreover, just by seeing this pattern we might have been tempted to think that it extended indefinitely (especially because checking it for higher values of n gets increasingly difficult), which would have been incorrect. Experimentation, in other words, is important but having only a small amount of it can potentially mislead us. The lesson for many people, physicists in particular, is to avoid jumping to conclusions too early and keep on checking!

19

$$\binom{n}{k} = \frac{n!}{k!(n-k)!}$$

One might still wonder, however, whether there is another explanation for why this problem led us awry. Consider the identity

$$1 + \binom{n}{2} + \binom{n}{4} + \binom{n}{6} + \dots \binom{n}{n} = 2^{n-1}$$

(for odd-numbered n's, the last term is $\binom{n}{n-1}$)). You can prove this by thinking about the binomial expansion of $(1+1)^n = 2^n$ and $(1-1)^n = 0$. Up to $n = 5$, the above formula agrees with the answer we have, and so that explains why we get the exponential answer for small $n < 6$. But for higher n, since the higher binomials are missing, the answers beyond that begin to deviate from our early expectations. In short, we not only succeeded in figuring out the right answer to this problem, we were also able to look back and explain how, why, and where we went amiss.

Puzzle

How many regions to you get on the plane if you draw $n \leq 3$ generic lines? How about the same question in three dimensions: How many regions do you get if you consider $n \leq 4$ generic planes? More generally, consider $n \leq d+1$ generic hyperplanes in d-dimensions–hyperplanes being planes of one dimension lower than the space in which they reside. How many connected regions will they divide this space into?

Solution

The answer is

$$\sum_{i=0}^{\min(d,n)} \binom{n}{i}.$$

Let's try to get some intuition about this answer for small values of d. If $d = 2$, a hyperplane is nothing but a (one-dimensional) line. A line divides R^2 into 2 regions, 2 lines divide it into 4 regions, 3 lines, however, only divide it into 7 regions, not 8. If $d = 3$, a hyperplane is an ordinary (two-dimensional) plane. A plane divides R^3 it into 2 regions, 2 planes divide it into 4, 3 planes divide it into 8, but 4 planes divide it into 15 regions, not 16. In the limit $d \to \infty$, it is just 2^n for all n. For finite d, it will be 2^n only for $n \leq d$.

This is a phenomenon that happens a lot in physics. There are many expressions that simplify at various limits. In this case, simplification occurs at the limit, $d \to \infty$.

The Paradoxes of Infinity

Infinity is a concept that has both intrigued and confused humanity over the millennia. The ancient Greek philosopher, Zeno of Elea, devised a series of paradoxes about infinity–at least nine of which are known to this day–which lead to seemingly absurd results. More than 2,400 years later, the notion of infinity still holds many mysteries for us.

The set of positive integers or natural numbers $\mathbb{N}$, as is well known, is infinite. Between any two natural numbers on the real axis, there are infinitely many rational numbers (which are expressed as the quotient of two integers). So our intuition might suggest, perhaps reasonably enough, that one of them–namely rational numbers–is more infinite than the other–integers. In that case, our intuition would be wrong, because there is a one-to-one correspondence, or "bijection," between the integers and the rational numbers. The total number or "cardinality" of the rational numbers $\mathbb{Q}$ is the same as that of $\mathbb{N}$.

One way to picture this is to represent rational numbers $\frac{p}{q}$ on a lattice and "wrap" positive integers around it like a spiral (Fig.50). We see that if we start counting points on the lattice as we spiral around, and skip points that are undefined (when $q = 0$) or unsimplified (when p and q are multiples of an integer), we get a bijection between $\mathbb{Q}$ and $\mathbb{N}$.

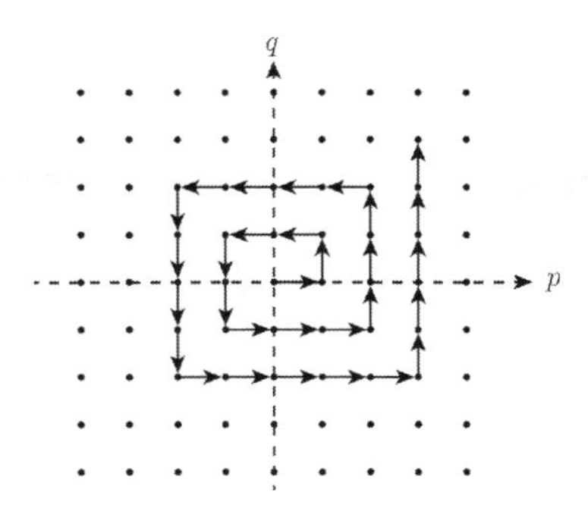

FIGURE 50. There are as many integers as rational numbers. We can see this by wrapping the positive integers and covering every pair of integers by spiraling around the plane.

Now consider the sets $\mathbb{R}$, the real numbers, and $\mathbb{R}^2$, the points on a plane. They are both obviously infinite, but the second would seem to be clearly bigger than the first. Nevertheless, they actually have the same cardinality. One way to write down a bijection is to take the decimal expansion of $x = x_1x_2x_3\ldots$ and form a pair of real numbers $x_1x_3x_5\ldots$ and $x_2x_4x_6\ldots$. This isn't quite complete yet, but it captures the idea.

There are more real numbers than integers, but are there any sets that have cardinality in between? This is related to the continuum hypothesis, which states that there are no such sets. However, this hypothesis cannot be proved or disproved. In other words, we can choose to add this fact (or its negation) as an axiom and still have a working system of mathematics without knowing for certain whether the hypothesis is true or not.

There is a relevant theorem in logic, known as Gödel's incompleteness theorem, which says that in *any* system of logic there are questions that cannot be proved true or false within that framework. A physicist might worry that the laws of the universe that one discovers might, therefore, never be complete. It is possible, according to Gödel's incompleteness theorem, that neither physical reality, nor the validity of certain physical phenomena, can be established from a finite set of axioms or laws. This is not yet a pressing issue in contemporary physics, but it may become one in the future as our theories become more mature and carry us closer to the frontiers of truth!

Hilbert's Hotel Problem

Another famous example related to infinity is the thought problem involving Hilbert's Hotel.

Puzzle

A hotel has infinitely many rooms indexed by the natural numbers $1, 2, 3, \ldots$. A traveler asks for a room but is told that all rooms are full. Can the traveler suggest a solution? What if a (countably) infinite number of travelers show up at once. Is there a solution to accommodate all of them?

Solution

Yes! If one traveler shows up she suggests that each person move over one room to the right $n \to n+1$, and suddenly there is a vacant room, namely room number 1. And if a countably infinite number of people show up at the hotel, we can have guests in the hotel go to double their room number, leaving a countably infinite number of odd-numbered rooms still vacant!

Surprisingly, this kind of paradox has applications in physics! Recall Dirac's example, which had infinitely many electrons filling the negative energy states–the "Dirac sea." It's conceivable that each electron could simultaneously shift up one energy level (Fig.51). In quantum mechanics, one can actually *create* electrons by moving up the energy levels and inducing this behavior. Similarly, if all the electrons move down by one level, we are left with a "positively charged" hole–a positron. We should seriously consider mathematical pathologies–or paradoxes concerning infinity that may not be fundamentally unsound–because sometimes they can actually model reality!

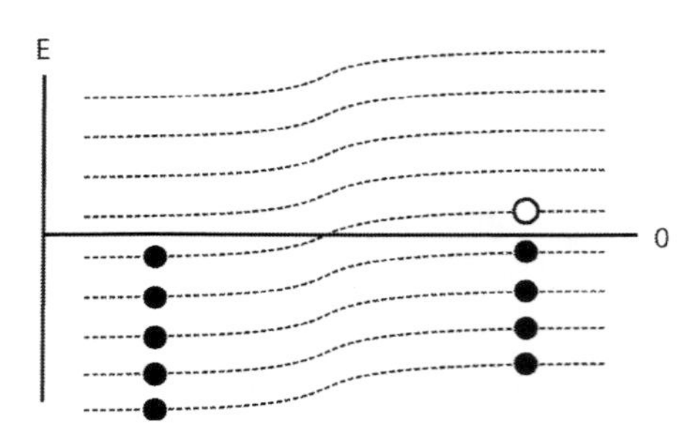

FIGURE 51. After an upward energy shift, the states in the Dirac sea move one step upward and in the process we have created an electron!

Puzzle

(Counter-intuitive math) Suppose that you take two different kinds of pills (A and B) a day, one of each kind. The pills are indistinguishable to the eye. One day, you accidentally mixed together two pills of B and one of A. How do you take the medicine without wasting these three pills?

Solution

Take an A pill and put it with the three questionable pills. Then proceed to split each pill in half, carefully separating the halves into two distinct piles. Then we are left with two piles, each with the correct daily dose.

Puzzle

You are at a party where 5 couples comprise all the people at the party. Everyone shakes hands only with people they don't already know. Everyone other than your spouse tells you that they shook a different number of hands. How many hands did your partner shake?

Solution

Everyone can shake hands with up to 8 people. (They already know their spouses and themselves). There are 9 possibilities for the number of handshakes and 10 people altogether, so your partner must have shook the same number of hands as someone else.

One person of the 4 couples must have shook 8 hands and one must have shook 0 hands. The person shaking hands with 8 has shook hands with everyone except his spouse. So that person must be married to the person who shook 0 hands. Similarly we can pair off 7,1 and 6,2 and 5,3. Note that all of the people alluded to above shook a different number of hands. That leaves one remaining possibility: You and your spouse must have each shaken 4 hands.

Analytic Series

Mathematicians are interested in trying to make sense of this kind of puzzle, which often comes up in physics too. As a teaser, let us prove that

$$1 + 2 + 2^2 + 2^3 + 2^4 + \cdots = -1.$$

Let $x = 1 + 2 + 2^2 + 2^3 + 2^4 + \cdots$. Then it is easy to see that if we multiply it by 2 and add 1 you get the same series again. In other words we can simply write

$$2x + 1 = x,$$

which implies that $x = -1$. This is a counter-intuitive result that is, nevertheless, correct. While this simple derivation was somewhat lacking in rigor, we can be mathematically more precise by means of "analytic continuation." We recall the expansion of $(1-x)^{-1}$ which is related to geometric series

$$\frac{1}{1-x} = \sum_{n=0}^{\infty} x^n.$$

The right-hand side of the equation only makes sense for $|x| < 1$. But the left-hand side makes sense even if $|x| > 1$. We can use the left-hand side to make sense of what we mean by the right-hand side even if $|x| > 1$. This is called an analytic continuation of what the right-hand side represents for $|x| > 1$.

As another example, let us talk about the sum of all natural numbers

$$1 + 2 + 3 + 4 + \cdots = -\frac{1}{12}.$$

The fact that this is finite and negative is again counter-intuitive. We can define an analytic function, which is called Riemann's zeta function, as follows:

$$\zeta(s) = \sum_{n=1}^{\infty} n^{-s}.$$

This function is defined in the complex plane when s has real part > 1, but it can be analytically continued in the complex plane to a unique function, just as we did with the geometric series above. It turns out that once we do this, it satisfies $\zeta(-1) = -\frac{1}{12}$. If we naively substitute $s = -1$ in the defining expression, then one would be forced to conclude that

$$\sum_{n=1}^{\infty} n = -\frac{1}{12}.$$

Again, this kind of calculation crops up in physics (specifically in finding the number of dimensions that bosonic strings live in), which comes from the equation

$$(d-2)\left(\frac{1}{2}\sum_{n=1}^{\infty} n\right) = -1 \implies d = 26$$

where d is the number of dimensions. This is where the 26 dimensions of early versions of string theory come from.[20]

[20]Technical aside about superstrings and groups: Fermionic/superstrings live in a space-time dimension $d = 10$. This is special because $d - 2 = 8$. We will later see why 8 dimensions are so special. One thing that has turned out to be crucial in physics is the mathematical notion of a *group*. These encompass symmetries. Let us first talk about discrete groups, which can be thought of as consisting of symmetries like reflections and discrete relations. For instance, the symmetries of a square form a group.

In physics, singularities pop up here and there, which we are not well-equipped to deal with. But we can go past the singularity and find that there are still well-defined points beyond it. This is why ideas from complex analysis are helpful for theoretical physics. When we encounter this kind of infinite series, we try to proceed analytically, which means we try to come up with functions, such as the zeta function, that make sense even in regions where one would not expect them to. Sometimes the way to proceed analytically may not be unique, but if several methods gives us the same answer, then we can be more confident that we can use it in a physical theory.

Similarly, the symmetries of any regular polygon form a discrete group. Are there analogous discrete symmetries of 3-dimensional objects? The group SO(3) consists of 3×3 matrices M satisfying $M^t M = I$, where I is the identity. These matrices have the property that they preserve *lengths*. More precisely, the dot product on $\mathbb{R}^3$ is given by $\langle w, v \rangle = w^t v$. If $M \in \mathrm{SO}(3)$, then $\langle Mw, Mv \rangle = w^t M^t M v = w^t v$. You can similarly talk about "rotations" in 8 dimensions and define the group SO(8) to be the 8×8 matrices M satisfying $M^t M = I$. The number 8 here is very special; there is a phenomenon that does not occur in any other dimension. In d dimensions, there are "spinors" which have roughly $2^{d/2-1}$ for even dimensions. Now, SO(d) acts on d-dimensional vectors. Usually, $2^{d/2-1} \neq d$, but when $d = 8$ something special happens, and they agree! This coincidence turns out to be directly related to why superstring theorists think the world has dimension d satisfying $d - 2 = 8$ leading to $d = 10$ dimensions. This is deeply related to the existence of supersymmetry in superstring theory.

Puzzle

Let us consider something familiar–a piece of standard, letter-sized paper. You have a pair of scissors, and you want to cut the paper in such a way that it remains connected, but your whole body can pass through it. Is this possible?

Solution

You might assume that this is obviously not possible for the paper is only so big and there's no way to make it much larger. Yet the assigned task is indeed doable, as you can see by cutting the paper along the lines indicated below.

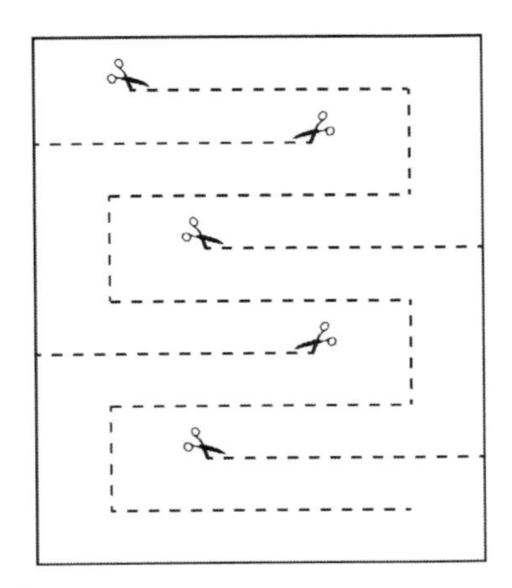

Those who found this result surprising, and contrary to their intuition, may have been confused by the fact that area and perimeter are different things whose scales do not have to be closely aligned. This is similar to the way that some of us might have been tricked by the problem involving the belt wrapped around the equator.

Puzzle

There are 100 coins scattered in a dark room. 90 have heads facing up, and 10 have tails facing up. You cannot distinguish (by feel, etc.) which coins are which. How do you sort the coins into two piles that contain the same number of tails?

Solution

Note that the piles need not be the same size! Choose any 10 coins to make one pile, then turn over all 10 coins in that pile. The number of tails in this pile is now equal to the number of tails in the other pile of 90 coins!

Perhaps you're surprised to hear that this problem, which might have sounded difficult, could have had such a simple solution. To see that it is true, suppose there were x tails in the 10-coin pile. Since there were initially a total of 10 tails, there must be $10 - x$ tails in the pile of 90 coins. When you flip over all 10 coins in the smaller pile, the x tails in that pile become heads and the remaining coins, which must be $10 - x$ heads, become tails. That matches the larger pile. So the solution is, in fact, correct, and all it required was some elementary math.

Puzzle

Consider the curve $y = 1/x$ from $x - 1$ to $x = \infty$. If you rotate this around the x-axis, you get a funnel-like surface of revolution known as Gabriel's horn (Fig.52). How much water can you fit inside this surface? And can you "paint" this surface?

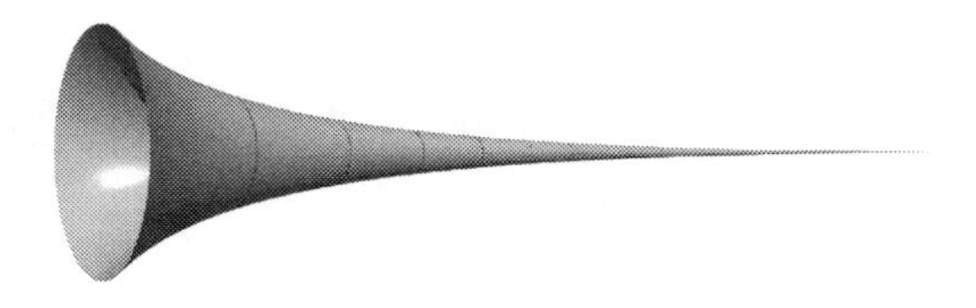

FIGURE 52. Gabriel's horn with finite volume but infinite area

Solution

It is impossible to paint the surface because it has *infinite* area, even though it has a *finite* volume and can only hold a finite amount of water. You can calculate the area as

$$A = \int_1^\infty \frac{2\pi\sqrt{1 + (y')^2}}{x}\, dx = \infty.$$

However, the *volume* is finite:

$$V = \int_1^\infty \frac{\pi}{x^2}\, dx = \pi.$$

This counter-intuitive puzzle challenges our notions of what we mean by "painting" something. Physically, it means to cover with some positive, uniform thickness of paint, but mathematically, the thickness goes to 0. So volume and area cannot really be compared. They are, as the saying goes, as different as apples and oranges.

The Monty Hall Paradox

Puzzle

You are participating in a game where the host has placed a prize in one of three covered boxes. You are asked to pick one of the three boxes. After picking it, but before opening the box, the host opens one of the other two boxes, which she knows carries no prize in it. She then gives you the option of switching your pick. Are you better off switching or not?

Solution

It is always better to switch, since your initial chance of picking correctly is $\frac{1}{3}$. So the chance of picking correctly after switching is $1 - \frac{1}{3} = \frac{2}{3}$.

Many people find this unintuitive, since it seems like one should have a $\frac{1}{2}$ chance. Part of this intuition stems from psychology: we are either reluctant to trust the game host or resistant to change. And part of this natural reaction is just plain wrong, constituting an abuse of probability theory. To make the situation more intuitive, let us imagine the game involved 100 boxes, instead of three. The host opens 98 other boxes, which she knows to be empty, after you first pick a box. Would you switch to the other leftover box? In this case, it should be obvious, even without a calculation, that it is in your best interest to switch. After all, it is very unlikely that your first $\frac{1}{100}$pick was the right one!

Here's an absurd example of how psychology (and the specious application of probability) can mislead us: As the Large Hadron Collider (LHC) was being built at the CERN laboratory, someone claimed that CERN had a 50% chance of destroying the Earth by creating a black hole in the collider. That argument was based on the following spurious reasoning: Either the LHC will destroy the Earth or it won't, so it is a 50/50 chance either way.[21]

Puzzle

Can you draw a closed curve on the plane such that no square can be inscribed in it?

Solution

This is believed to be possible (according to the *Toeplitz conjecture*), but it is not known. Otto Toeplitz proposed this problem in 1911 and proved the special case of it (when the curve

[21]Walter L. Wagner was quoted as saying this on *The Daily Show*. Wagner filed a lawsuit *Sancho v. U.S. Department of Energy, et al.* that was dismissed (in an apparent vote for reason). See `http://www.nytimes.com/2008/03/29/science/29collider.html`

is piece-wise the graph of a smooth function)–in other words, for a non-self-intersecting, continuous closed curve–then we can always inscribe a square. At first, this statement may sound very unintuitive, but it is not too hard to argue based on the choices one has–selecting the first point of the square and then deciding the length of square and the orientation angle needed to fit the other three vertices–that there are are enough degrees of freedom to satisfy this conjecture (Fig.53).

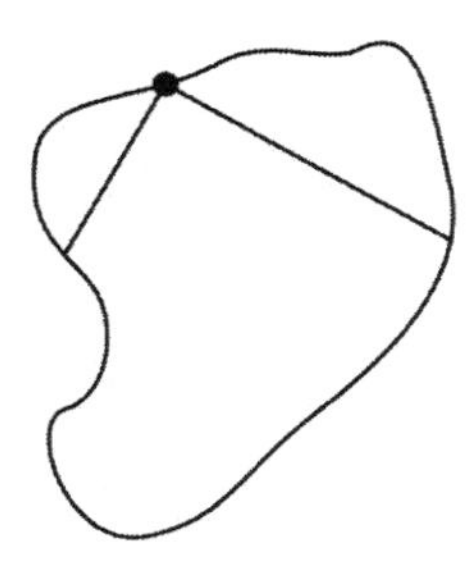

(A) First we choose a point on the curve and draw two perpendicular line segments.

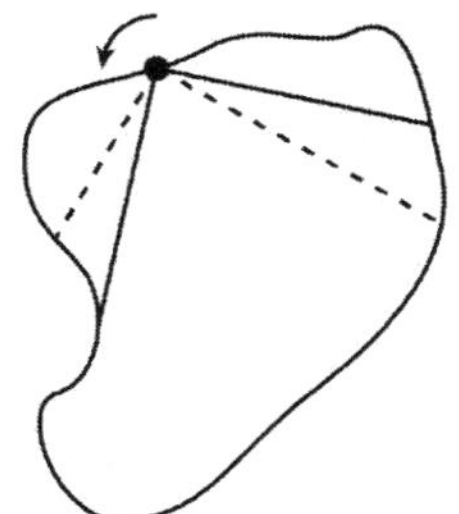

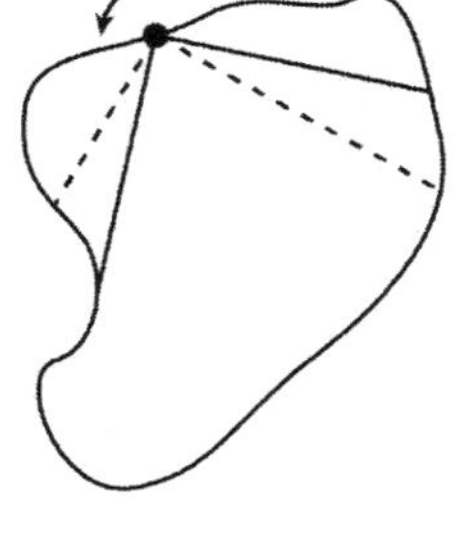

(B) We can rotate these line segments until they become equal in length.

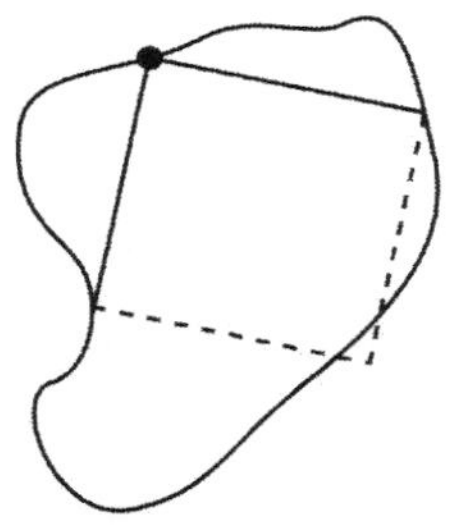

(C) Then we complete the square defined by those line segments. Alas, the last point does not land on the curve.

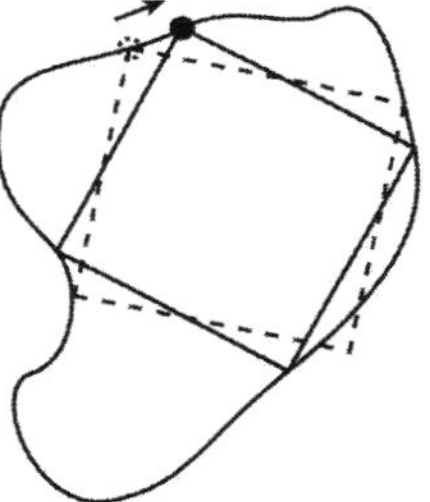

(D) We can repeat the previous steps for all points on the curve. Eventually we will find an inscribed square.

FIGURE 53. The special case of the Toeplitz conjecture: every plane smooth closed curve, contains all four vertices of an inscribed square.

Logical puzzles are often very counter-intuitive. Here is an example.

Puzzle

A man who has been sentenced to death is told that he will be executed sometime next week between Monday and Friday. He is also assured by law that he will not be able to know the day on which he will be executed, but that he would be informed by 10 a.m. on the day that he was going to be executed. The man deduced that he could not be executed. Can you explain his reasoning?

Solution

If he were to be executed on the last day, then he would be able to deduce it on Thursday, by observation of the fact that it was the last day on which he could be executed. Since this is against what was promised to him, he knows he will not be executed on Friday. But then, by the same inductive logic, he could not be executed on the penultimate day either. Continuing this logic day by day, he deduces he cannot be executed at all, as long as the promise made to him is upheld.

Satisfied with his reasoning, the man was confident that he could not die. However, the prisoner was executed on Tuesday, and he did not know a day in advance. Nor could he have predicted the day of his execution. The conditions he was promised were thus satisfied, and the man met a somewhat unpredictable though still expected end.

Here is another example: consider the complex numbers. One cannot find a real number solution to $x^2 = -1$, so one simply "creates" a new number i such that $i^2 = -1$ by definition! We can define complex numbers as pairs of real numbers with the multiplication rule

$$(x_1, y_1) \cdot (x_2, y_2) = (x_1 x_2 - y_1 y_2, x_1 y_2 + x_2 y_1)$$

If we try to solve for $(x, y) \cdot (x, y) = (-1, 0)$, then we get $x = 0$ and $y = 1$. You might object that this is cheating, but it often turns out that new ideas in math involve unintuitive constructions. The construction of 0, negative numbers, fractions, and real numbers have a similar history. Complex numbers, moreover, turn out to be fundamental to modern physics, particularly in quantum mechanics.[22]

Here is yet another example. Orientable surfaces can be drawn in three dimensions, and they are always spheres with g-holes. In an orientable surface, the coordinate system (which in the 2 dimensional case consist of, say, x,y axes) does not change as you move about the space. However, the axes will flip at some point, or points, when you move around a non-orientable

[22]How does this get used in physics? What does it mean to have a complex number of apples, say? Well, it turns out that physical quantities will always be measured as real numbers, even in quantum mechanical formulations. But complex numbers are still essential to the formalism of quantum mechanics.

surface. Furthermore, there are some non-orientable surfaces that cannot be placed in three-dimensional space. One such surface is a "Klein bottle." To describe this, let us first consider constructing a torus by rolling up a square piece of paper into a cylinder and then attaching the edges of the cylinder together (Fig.54). For the second step of gluing we can instead to the following: We flip the orientation of the edges when we glue them. We get what is called a "Klein bottle" rather than a torus. An important lesson to draw from this is that things cannot always be nicely embedded (without self-crossing) in low-dimensional space (Fig.55).[23]

After toying around with these objects for a while, one can get more accustomed to thinking about higher dimensions. For instance, if you have a 7-dimensional plane and an 8-dimensional plane in a 10-dimensional space, then they will generically intersect in a 5-dimensional space. You can reason that this is so by thinking, analogously, about lower-dimensional space. In 2 dimensions, two 1 dimensional lines typically intersect at a 0-dimensional point ($1 + 1 - 2 = 0$). In 3 dimensions, a 1-dimensional line and a 2-dimensional plane generically intersect at a 0-dimensional point ($0 = (1 + 2) - 3$) and, similarly, two 2-dimensional planes intersect at a 1-dimensional line in three dimensions ($1 = (2 + 2) - 3$). By analogy, we can deduce that

[23] figure made by Vierkantswortel2 on Wikimedia Commons.

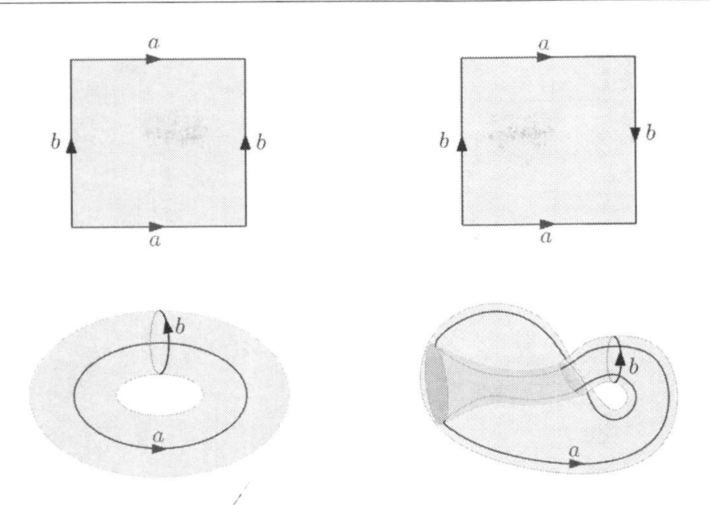

FIGURE 54. Depending on how we glue the opposite sides of a rectangle we can obtain a torus (as in left figure) or a Klein bottle (as in the right figure).

the answer to the original question is $(7+8)-10=5$. The point is that by applying our reasoning to familiar objects, we might be able to deduce results about much more exotic things–things like 7- and 8-dimensional planes that we can't even picture in our mind. But at the end of the day, of course, you should always check your intuition with rigorous mathematics.

Puzzle

For answering a true or false question, Jill has a 1 in 10 chance of getting it right, and John has a 7 in 10 chance of getting it right. Who would you pick to help you answer the question?

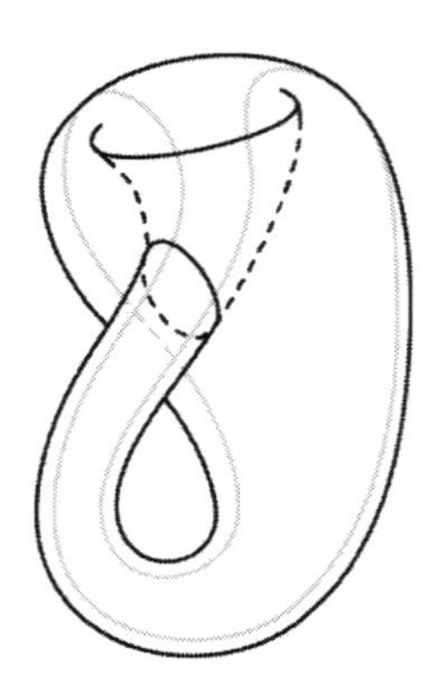

FIGURE 55. A Klein bottle embedded (as best as anyone can) in 3-dimensional space. While this embedding is problematic due to self-crossing, the bottle, itself, is mathematically sound.

Solution

The best strategy is to pick Jill, but to negate her answer by doing the opposite of what she says. Then you will have a 9 in 10 chance of getting the correct answer.

Puzzle

A group of people with assorted eye colors live on an island. They are all perfect logicians and if a conclusion can be logically deduced, they will do it instantly. No one knows the color of their eyes. Every night at midnight, a ferry stops at the island. Any islanders who have figured out the color of their own eyes then leave the island, and the rest stay. Everyone can see everyone else at all times and keeps a count of the number of people they see with each eye color (excluding themselves), but they cannot otherwise communicate. Everyone on the island knows all the rules, as stated in this paragraph.

On this island there are 100 blue-eyed people, 100 brown-eyed people, and the Guru (she happens to have green eyes). So any given blue-eyed person can see 100 people with brown eyes and 99 people with blue eyes (and one with green), but that does not tell him his own eye color; as far as he knows the totals could be 101 brown and 99 blue. Or 100 brown, 99 blue, and he could have red eyes.

The Guru is allowed to speak once (let's say at noon), on one day out of all their endless years on the island. Standing before the islanders, she says the following:

"I can see blue eyes."

Who leaves the island, and on what night?

Solution

The answer is that on the 100th day, all 100 blue-eyed people will leave!

If you consider the case of just one blue-eyed person on the island, you can show that he obviously leaves the first night, because he knows he's the only one the Guru could be talking about. He looks around and sees no one else with blue eyes and, therefore, knows that he should leave.

If there are two blue-eyed people, they will each look at the other. They will each realize that "If I don't have blue eyes, then that guy is the only blue-eyed person. And if he's the only person with blue eyes, then he will leave tonight." They each wait and see, and when neither of them leaves the first night, each realizes "I must have blue eyes." And each leaves the second night.

This induction can continue all the way up to day 99, when everybody will know that they have blue eyes. Thus they will each wait 99 days, see that the rest of the group hasn't gone anywhere, and on the 100th night, they all leave.

Puzzle

This birthday problem is a classic unintuitive problem: What is the probability that there is at least one shared birthday in a group of n people?

Solution

When n is 23, the probability is about 50%. When n is 50, the probability is about 97%. This is a surprisingly high probability. It follows because the probability of no two people having the same birthday is

$$\frac{365 * 364 * ... * (365 - n + 1)}{365^n}$$

and so the probability that at least two have the same birthday is

$$1 - \frac{365 * 364 * ... * (365 - n + 1)}{365^n}.$$

Much of probability can be unintuitive. But the situation is not hopeless; with time and practice, problems like those we have just discussed can become more familiar and, eventually, more in line with our intuition. Building up intuition is a key, not only in mathematics but also in physics. And that is exactly what we will be looking at next.

6. **Physical Intuition**

Intuitive Physics

After our foray into counter-intuitive math, we are back to studying the role of intuition in physics. A lot of physical intuition is ingrained, but it can also be cultivated. Physicists would love to develop their physical intuition to the point where they can quickly answer physical questions without needing to do detailed calculations, using explicit calculations only to refine and quantify their intuitive understanding.

Richard Feynman was one such physicist, famous for having very good physical intuition. But it wasn't as if he was simply born with great intuition. Much of that came to him as a result of the detailed mathematical calculations he had carried out. Afterwards, he would often go back and ask if he could have foreseen the result without having done the calculations to begin with. In many cases, he found simple intuitive explanations for results that were originally developed through difficult mathematical computations. So, intuition did not come for free. But the next time Feynman encountered a similar problem, he did not have to go through all that hard work. He could use his intuition to "guess" the answers to some nontrivial questions.

It is important to keep in mind that intuitive physics is not trivial physics. It is only trivial if you look at a problem in exactly the right way. The same thing can happen with good puzzles too. You can work on finding the solution to a puzzle for a long time, without having a clue about how to proceed. The problem may seem impossible to solve until something switches in the brain, and you recognize a new way of approaching the task that suddenly makes it seem much more manageable. The mental reorientation required to solve the problem is generally rather nontrivial, even though the solution itself–after the "switch" has been flipped–may be quite simple. That is why doing puzzles can be great practice for doing physics, and vice versa. Strategies that are effective in the one realm may prove fruitful in the other realm as well.

Galileo Galilei

When Galileo di Vincenzo Bonaulti de Galilei was studying the laws of motion, there had already been considerable philosophical discussion on that subject. Aristotle, for instance, had put forward the intuitive idea that heavier objects accelerate faster during free fall. This is, of course, a notion that many of us share anyway. Galileo refuted that proposition by means of an experiment, by famously dropping small and large objects off the Pisa tower, showing that lighter and heavier objects accelerate at the same rate during free fall. Many people found this surprising. The experimental method, generally speaking, was not held in high regard in that era. People were more inclined to accept facts that were arrived at through "pure reason," believing that one should not have to soil one's hands by performing experiments in order to attain the truth.

Afterwards, Galileo also provided support for his result through pure reasoning, and the logic he employed was so elegant and simple it made the outcome of his experiments seem obvious. The argument went roughly as follows: Suppose you take two objects of exactly the same shape and mass.

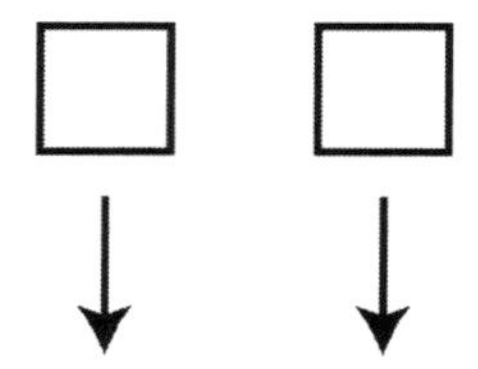

You release both at rest from the same height. Which one will hit the ground first? Obviously, since the objects are the same size, they should hit the ground at the same time. This is simply a statement of translational symmetry in the horizontal direction. Now take a third object of the same size and drop it alongside the original two, which are again released from the same height at the same time.

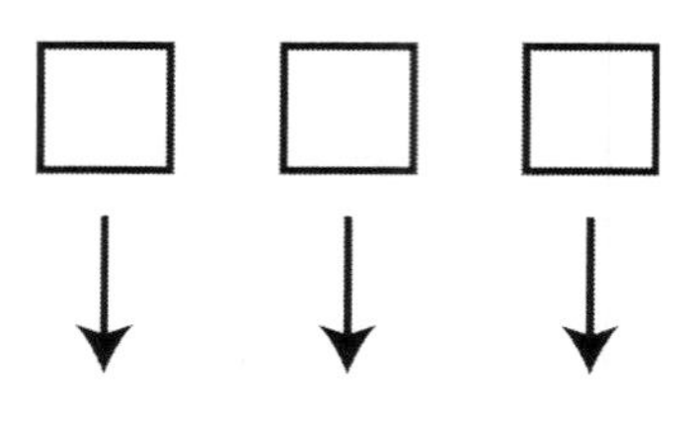

Again, it is clear that they will hit the ground at the same time. It also seems apparent that shifting the three objects horizontally before releasing them will not change the fact that all three will still hit the ground at the same time. Now imagine moving the original pair close to each other, so close that they appear to be the same object.

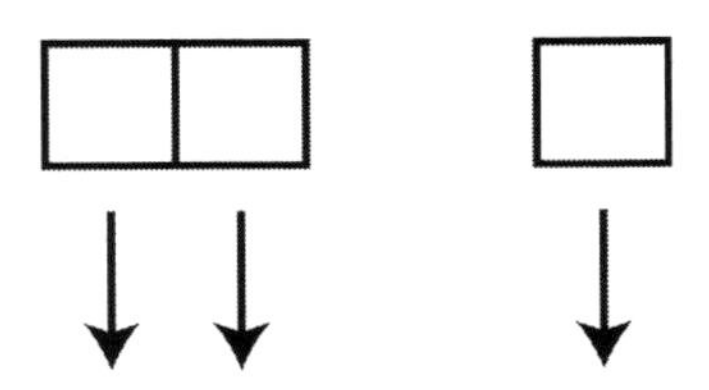

They might as well be considered as a single object of twice the size, but obviously they *still* hit the ground at the same time as the other object. So the object with twice the mass of the other one hits the ground at the same time. Now the result does appear *obvious*! This level of clarity is not often achieved, suggesting that our physical intuition needs to be refined to give us the correct results. Our original intuition that heavier objects hit the ground first may be rooted in psychological preconditioning: Perhaps we pay more attention to the heavier object because it can impart more energy upon impact than lighter objects can. And if we are naturally prone to give weightier objects more attention, we might as well assume that they land first, too.

Isaac Newton

The famous story goes that Sir Isaac Newton's ideas about gravity were catalyzed by an apple falling from a tree in his mother's garden. He observed its descent in some versions of the tale; in others he was struck on the head. In either case, the falling apple purportedly made a deep impression on him. But what, one might ask, does an apple have to do with the orbits of planets as governed by gravity?

This is how Newton described it in his book, *Philosophiae Naturalis Principia Mathematica.* He asked, why doesn't the moon fall like an apple? To get a handle on this question, he modified it somewhat. Suppose there is a tall mountain on the North Pole, and we fire a cannonball from the top of the mountain as in Fig.56. It will fall down to the ground, of course. But suppose you fire the cannonball with more force. Then it will still fall to the ground, though landing much farther away in this instance. And if you fire it really hard, it will move a distance comparable to the size of the Earth, perhaps falling on the equator. If you fire it even harder, it falls even farther, perhaps landing on the South Pole. Suppose you now give it an even stronger kick. Then the cannonball will miss hitting the Earth completely and come all the way back around.[24]

Now we can compare the moon to the cannonball from our

[24]We are ignoring the air resistance in the discussion above.

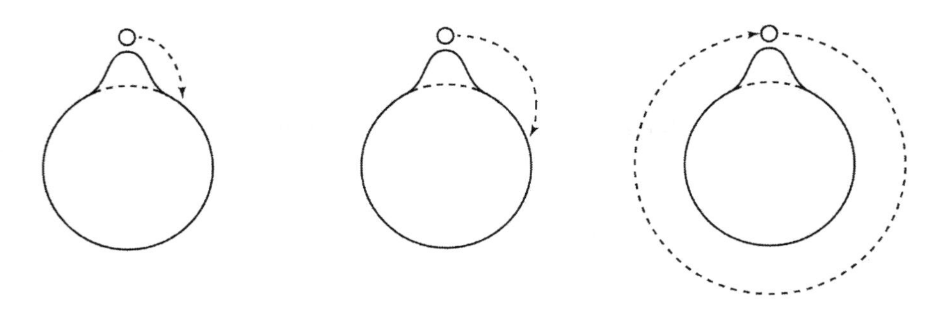

FIGURE 56. A falling cannon ball with a high enough initial velocity behaves very much like an orbiting moon.

thought experiment, and the reason it is not falling becomes obvious: Were it not for the Earth, the moon would have flown straight out into space. But because of the Earth's gravitational pull, the moon deviates from that path. It keeps falling towards Earth, but it also keeps missing Earth, because Earth is round. And that is what we end up with: a cannonball going around the Earth on an orbit. So indeed the moon is falling towards the Earth but the Earth is round and so the moon keeps missing it as it tries to fall on it! Earth's roundness is what prevents the moon from getting any closer to it.

Now, you can also intuit the fact that launching an object at a critical velocity could send it into a circular orbit. If the object moves too slowly, it will drop down. If the object moves too fast, it will head straight out into space. If it goes not quite so fast, it might then enter a circular orbit. However, we cannot conclude from the foregoing discussion that the orbit has to be periodic. Calculations are needed to establish that fact, though intuition can still carry us far.

Originally, the issue of the moon falling was not intuitive. But after changing our point of view, it made more sense–perhaps to the point of becoming obvious.

Puzzle

A truck moving slowly at 10 mph collides with a fly going in the opposite direction at 20 mph. The fly sticks to the window of the truck after collision. What is your guess about the speed of the truck after collision? (Do not use any equations).

Solution

We know well from physical intuition that the speed of the truck is barely changed. Needless to say, precise mathematical computations based on physical laws confirm this intuition.

Physical Intuition in Mathematics

Physical intuition can also be used to deduce mathematical truths. We will touch on a few such examples in this section.[25]

[25]Many of the examples discussed in this section are borrowed from the beautiful book, *The Mathematical Mechanic: Using Physical Reasoning to Solve Problems*, by Mark Levi. Readers seeking more illustrations of the ways in which physical intuition can influence mathematics are encouraged to consult this book.

Let us revisit Torricelli's theorem, which was briefly alluded to (in a footnote) in Chapter 3 (as part of finding the shortest highway system for four cities on the corners of the square): Suppose we have three points on a table and want to connect them in such a way as to minimize the sum of the distances between them. We argued mathematically that the path must form a "trivalent" graph–a collection of three lines meeting at a vertex–with three angles of 120°. Here is a physics-motivated way of seeing this: Imagine cutting holes in the table at those same three points and then attaching three identical balls to the tips of three strands of string (of fixed length), connected at a common vertex above (Fig.57). Assuming the masses of the balls are equal to m, the potential energy of the total system is $-mg$ times the sum of the lengths of the strings hanging beneath the table. This is minimized when the lengths of the strings hanging beneath the table are maximum. At such a point, the length of the connecting strings on the table is minimized, because the total length of the strings is fixed. By applying the equilibrium condition, we see that the tensions at the vertex must balance, and since the tensions are all equal (being pulled by equal weights), the angles must be equal too and, hence, must be 120°. This shows that bringing physics to bear on what was originally a mathematics problem can give us insights and lead to simple solutions that appeared to be difficult in the original mathematical formulation.

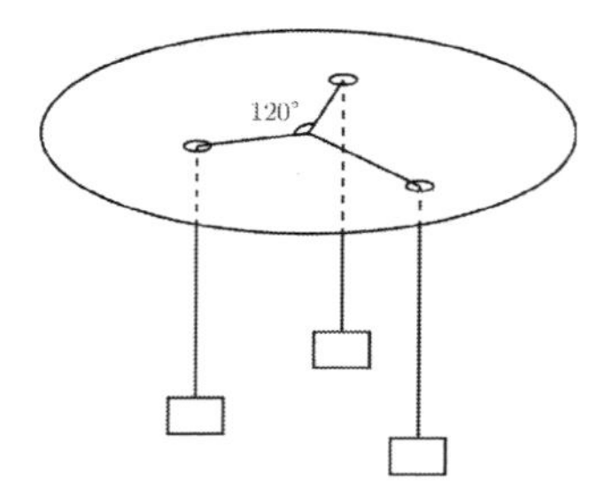

FIGURE 57. The strings attached to three equal weights settles at a point which minimizes the total length of the strings on the table. It is easy to see that the angles should be 120°'s based on balancing three equal magnitude forces pulling the common point in three different directions.

Line of Best Fit

Imagine a graph of data points (x_i, y_i). Suppose we want to find the line of best fit (for the purposes of linear regression), meaning that we want to minimize the sum of the squares of the vertical distance from the data points to the line. We can treat this as a physical situation by imagining that the points are nails on a table with the horizontal direction denoted as the x-axis and the vertical direction as the y-axis, as in Fig.58. The nails, in turn, are attached to springs that are connected to rings on a straight rod. The springs are confined to vertical tubes that allow them to move only in a vertical direction. The distance squared between the nails to the rod thus represents the potential energy of the springs, so finding the best linear fit is exactly the same as finding the configuration of the rod that minimizes the springs' potential energy.

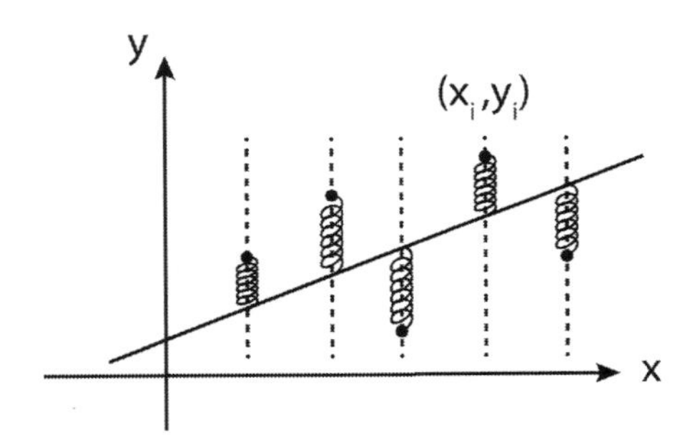

FIGURE 58. The line of best fit can be realized physically by where the rod, connected to data points through springs, settles.

Now let us solve the problem with physics. We know that in equilibrium, the forces and torque on the rod must balance. If the rod position is described by $y = mx + b$, then the force on the rod by the i-th spring is proportional to $y_i - (mx_i + b)$ (by Hooke's law), so the balancing of forces says that

$$\sum_i y_i - (mx_i + b) = 0$$

Now consider the torque on the rod. Each spring exerts the

torque, $x_i F_i$, so our second condition is

$$0 = \sum_i x_i F_i \implies \sum_i x_i \left(y_i - (mx_i + b) \right) = 0$$

Not surprisingly, these are exactly the same formulas as in the case for linear regression.

One may be unimpressed by the above model. All we did was to translate the math question to physics and solve it using math again. So what did we gain from all this shifting around? Here is the surprise: The physics modeling of this situation *prompts* a new question that might not have been otherwise apparent. To a physicist, it might seem artificial to limit the springs' motion to be strictly in the vertical direction. What if we were to remove this constraint and allow the springs to orient in any direction, as in Fig.59? After all, it seems natural to look for the configuration that minimizes the *distance* squared to the line, rather than just the vertical distance.

In fact, this would be the correct thing to do if the uncertainty in x and y were equally important. The usual linear regression presupposes that the uncertainty in y is much bigger than that in x. Thus, physical intuition allows us to reformulate mathematics problems in ways that can prove helpful–and might even yield new insights.

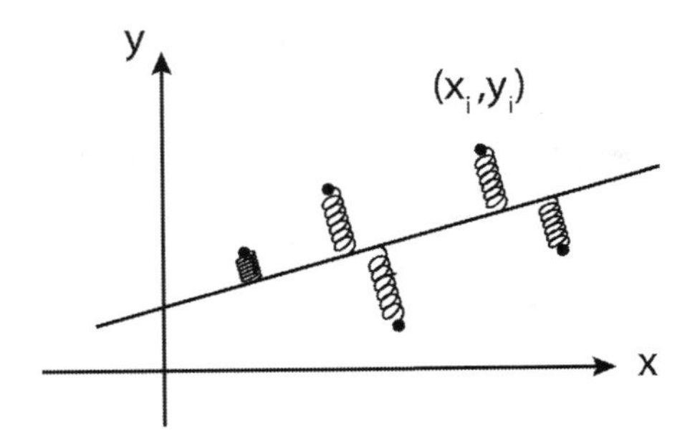

FIGURE 59. The springs not restricted to be along the vertical direction corresponds to the best line when both x and y have equal uncertainties.

Puzzle

Given a triangle and two medians of the triangle, is the line formed by the third vertex (and the intersection of medians (Fig.60)) also a median?

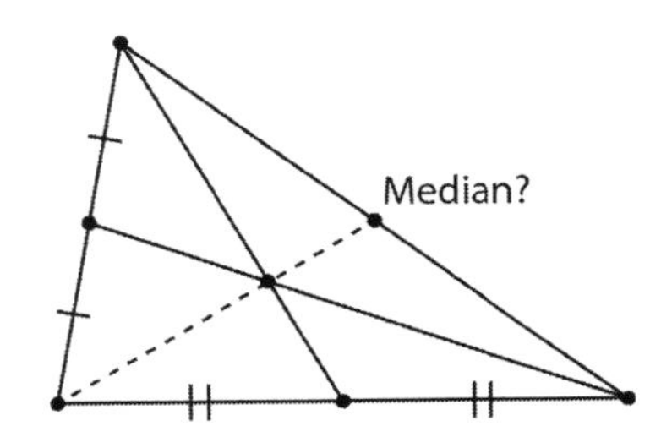

FIGURE 60. Do three medians of a triangle pass through the same point?

Solution

We will assume that the triangle is massless. Then we will bring physics into the picture by supposing that equal masses are attached to each vertex of the triangle as in Fig.61. If we find, through trial and error, the center of mass of this system of three masses, we can then balance it on a fulcrum, directly under the center of mass point.

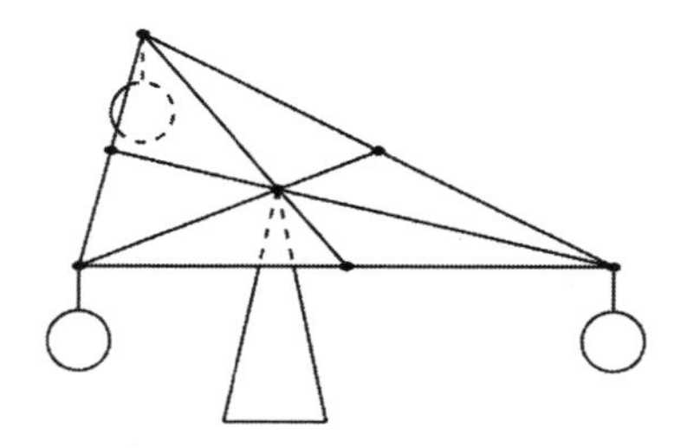

FIGURE 61. The medians should pass through center of mass in this physical model.

Now consider any two medians. The center of mass point must lie at their intersection point because for every median, the torques caused by the weights corresponding to the two other weights will balance and so the each median should pass through the center of mass. Since the object has one and only one center of mass, the third median must go through it too. Therefore, all three medians must pass through the same point.

Puzzle

Given a triangle of the following assigned lengths, what is the ratio of the lengths of the segments of the bottom side?

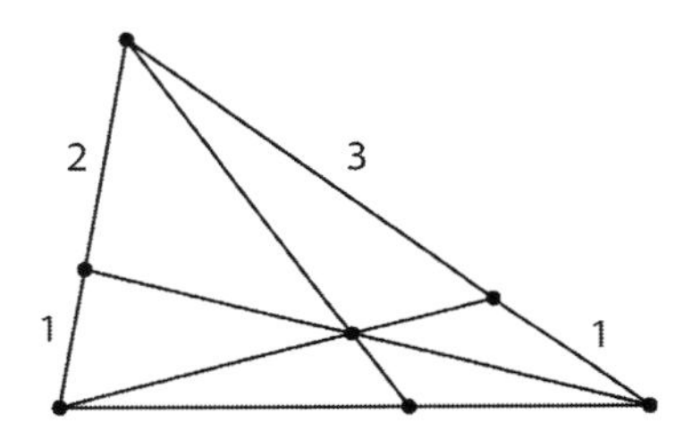

Solution

Let us place a fulcrum at the point of intersection between the lines that split the two sides by 2:1 and 3:1, as well as the line that splits the bottom into a yet-to-be-determined ratio. It can be discerned from looking at the figure that we need to attach unequal weights of $1, 2$, and 3 kg to the three vertices (1 at the top and $2, 3$ on left and right respectively) to balance the torques and triangle as a whole, thereby making this point of intersection the effective center of mass. From there, it is easy to see that the ratio of the segments along the bottom side must be 3 to 2. This exercise illustrates the power of physical intuition, as solving this problem without any physics is not an easy task.

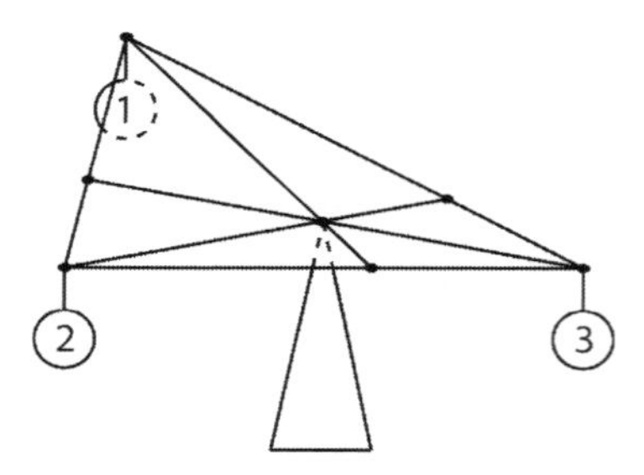

Puzzle

Consider positive real numbers. The AM-HM inequality says that the arithmetic mean is bigger than the harmonic mean:

$$\frac{a+b}{2} \geq \frac{2}{\frac{1}{a}+\frac{1}{b}}.$$

Show that the following (for which AM-HM is the special limit $a = d, b = c$) is true:

$$\frac{1}{\frac{1}{a+b}+\frac{1}{c+d}} \geq \frac{1}{\frac{1}{a}+\frac{1}{c}} + \frac{1}{\frac{1}{b}+\frac{1}{d}}.$$

Solution

We can use physics to show that this inequality is correct without having to carry out a complicated mathematical computation. What we have done in Fig.62 is to translate the problem into a electrical circuit diagram in which *a*, *b*, *c*, and *d* are resistors. The left-hand side of the inequality represents the resistance of the circuit as drawn here, without the switch flipped. The right-hand side represents the resistance after the switch is flipped. Resistance goes down whenever a free wire is added to a circuit, which is why this leads to an inequality. Based on our physical knowledge of how parallel circuits work, we know this statement is true without having to work through any complicated math. Physics leads to a visualization which helps with proving this otherwise purely algebraic problem!

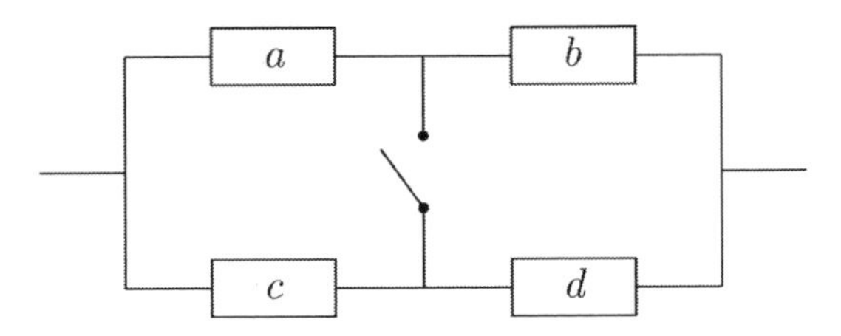

FIGURE 62. A resistor circuit designed to prove a mathematical inequality!

Archimedes' "Eureka"

Archimedes' principle of buoyancy holds that any object, wholly or partially immersed in a fluid, is buoyed up by a force equal to the weight of the fluid it has displaced.

This may sound unintuitive, but one can explain this principle in such a way as to make it seem almost *obvious*. Suppose we have an object, such as a crown, in a bucket of water. Now imagine performing some kind of "surgery" to remove the crown and put water back in its place, as in Fig.63. The rest of the water couldn't possibly tell whether water or a crown is there, so it acts the same way as if it were water. But with the crown replaced by water, the system is completely homogeneous, and we expect it to be in equilibrium. In other words, the water is not moving anywhere. This means that the net force from the rest of the water on that portion of the water must be equal to its weight, so that it could hold it up! When we put the crown back, *the rest of the water, again, could not possibly know* what is there. Therefore, the crown experiences the same buoyant force as if the water were there and so it experiences a reduction of its weight by the weight of the displaced water.

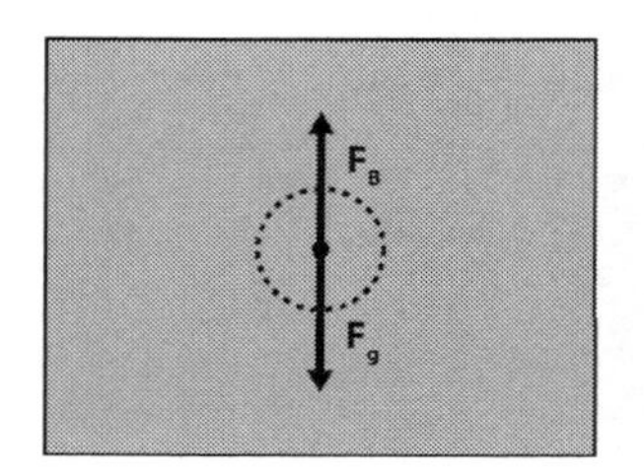

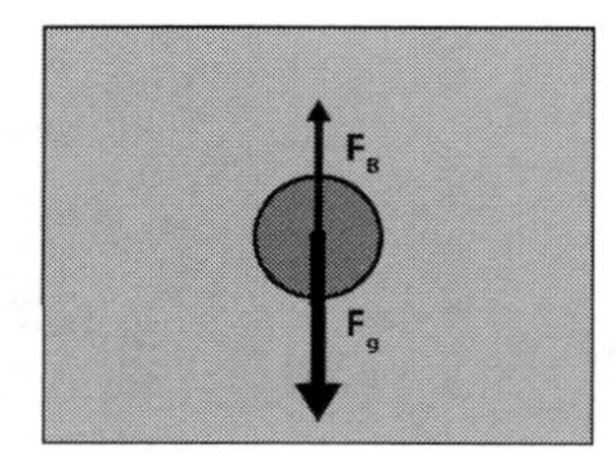

FIGURE 63. Since the rest of the water does not know what is there, it pushes up any immersed object as if it were water, by a force equal to the weight of displaced water.

By the way, the folklore surrounding Archimedes suggests that he came up with this law when thinking about how to identify a counterfeit crown. The accuracy of this story is not at all clear. According to the legend, he simply wanted to find the *density* of the crown. Since weight is relatively easy to compute, this is just a matter of measuring the volume of the crown, and that can be done by immersing it in water and observing the volume of displaced water. Perhaps the conservation of volume was not known before him and this discovery gave him a way to compute the volume and therefore density of an object by immersing it in water. So the buoyancy law is not required for this. Archimedes did, of course, discover the law of buoyancy.

Puzzle

Fill a box with small beans of some kind and bury a ping pong ball deep beneath. Shake the box for a while and observe what happens to the ping pong ball. Any explanations for what you see?

Solution

The ping pong ball floats to the surface for the same reason that less dense objects float to the surface in water. This is just another example of buoyancy: The ping pong ball is like an object immersed in fluid as in our previous example, and the "water" in this case is made up of small beans. The ping pong ball is less dense than the beans so the net force on it is upwards.

Pythagorean Theorem

As a demonstration of how physics can lead to new mathematics, we can now prove the Pythagorean theorem using physics. Consider a container shaped like a right triangle with side lengths a, b, c and filled with water (Fig.64). It seems intuitively clear that the container will not go anywhere; it must be in equilibrium. We will now explore what this statement of being in equilibrium means mathematically in terms of forces and torques. The force of the water on a particular side of the container is pressure (which is constant if the height of the container is small) times the area of that side, which is thus proportional to height times the side length. For the sake of simplicity (though without the loss of generality), we can normalize the height to be 1.

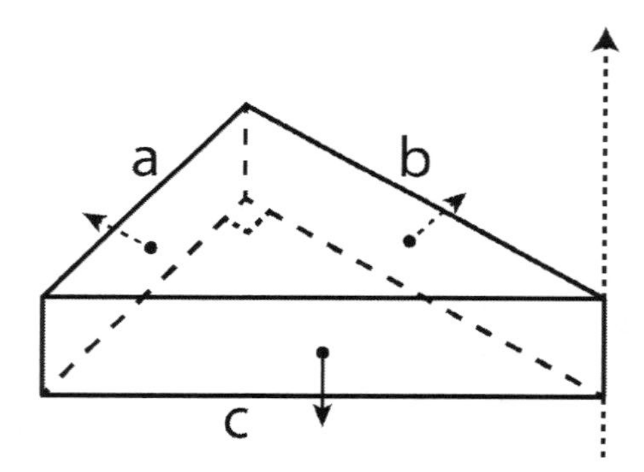

FIGURE 64. A right triangle shaped container filled with water can be used to demonstrate the Pythagorean theorem.

The fact that forces add up to zero is simply the statement that the three vectors form a triangle, i.e., the right triangle that we started with. Now let us take a look at the torque. We first need to choose an axis to measure the torque. Imagine putting a vertical rod through the vertex where the b and c edges meet. The net torque around this axis must be zero, otherwise the triangle would start spinning one way or the other. Recall that the torque is given by the product of the force and the lever arm. Therefore, the torque on the side with length b is $b \times \frac{b}{2}$. The torque on the side with length a is $a \times \frac{a}{2}$, pointing in the same (clockwise) direction, and the torque on the hypotenuse is $-c \times \frac{c}{2}$, since it is pointing in the other (counterclockwise) direction. Since this sum is zero, we deduce that $a^2 + b^2 = c^2$.

We can also derive additional mathematical statements: Imagine putting the rod through the vertex with the right angle. If $a < b$, then the torque relative to this axis is $\frac{a^2}{2} - \frac{b^2}{2} + c\Delta = 0$, where Δ is the distance between the midpoint of the hypotenuse and the point where the altitude meets the hypotenuse. Try proving this equality by geometric means alone, with no input from physics. In addition, if you are so inclined, you can also use the physics method to prove the more general result, the Law of Cosines, by considering a more general triangle.

It is not completely clear whether this can be viewed as an actual proof of Pythagorean theorem. The issue is whether, in our definition of torque, the Pythagorean theorem was somehow presupposed. At the very least, this is still an interesting way to think about the Pythagorean theorem in a physical setting.

Puzzle

A round table is supported on three legs placed every 120 degrees around the table. You want to tip the table over by exerting a downward vertical force somewhere on the perimeter. The question is, at what point along the circumference should you press down to minimize the amount of force needed to overturn the table? No calculation is necessary; just use your intuition.[26]

Solution

It is obvious that the ideal place is in the middle between any pair of the legs on the edge. Needless to say, one could use Newtonian mechanics to prove this statement. But if we arrived at any different answer, we would begin to question the assumptions leading to it. Nevertheless, if we want details, such as the amount of force needed to topple the table, we must do more precise calculations, which should, of course, confirm our intuition regarding where to exert the force. Sometimes, our intuition clashes with mathematics, and our intuition must be modified. However, sometimes we may make a mistake in applying the mathematics, in which case our intuition can serve as a powerful check.

[26]This problem was quoted in R.P. Feynman's lecture series.

Special Theory of Relativity

Intuition, oddly enough, can guide us towards conclusions that are counter-intuitive. For instance, Einstein used intuitive thought experiments while building a case for the special theory of relativity, which has many mind-boggling implications such as time dilation, length contraction, and the $E = mc^2$ formula.

Central to the special theory of relativity is one very unintuitive assumption: The speed of light is the same for everyone, no matter how fast one is moving relative to a light source. If one accepts this highly unintuitive principle, the rest of the theory's consequences become very intuitive.

Einstein was led to the concept of time dilation by a thought experiment. Imagine that you are sitting on a train, traveling along at a high velocity, $\vec{v}$, relative to the ground. It is a consequence of the special theory of relativity that if a person takes a train and returns to the point of origin, then his watch will become unsynchronized with that of a stationary passerby on the ground. Why?

Imagine a pair of mirrors in a car of the train, one on the ceiling and one on the floor separated by a distance L (Fig.65). Let us measure the time it takes a beam of light to bounce from mirror to mirror from the perspective of a person on the train, as opposed to the person on the ground. From the train passenger's perspective, the light is moving straight up and down, which is

$\Delta\tau = \frac{2L}{c}$ where c is the speed of light. However, from the bystander's perspective, the light is moving diagonally. If the train moves $v\Delta t$, then the time observed from the light beam to bounce back and forth is $\Delta t = 2\frac{L'}{c}$. Note that we are relying on the basic principle that the speed of light is the same speed, c, for both observers. But obviously $L' > L$, so $\Delta t > \Delta\tau$, i.e., time moves more slowly for the person on the train. Or, as is sometimes said, "moving clocks run slower."

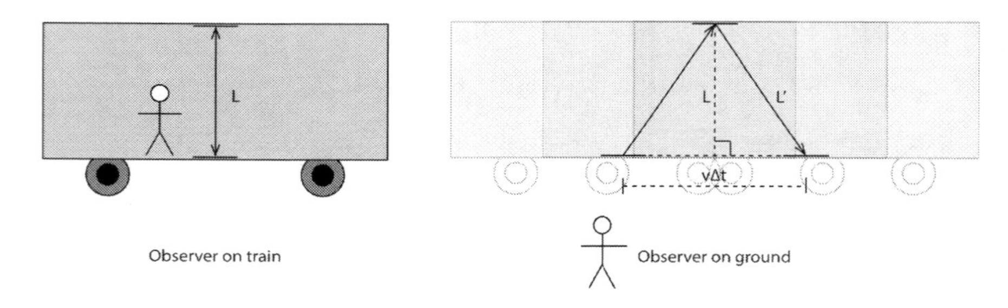

FIGURE 65. Time slows down on a moving train relative to the ground as can be seen by simply using Pythagorean theorem and assuming the speed of light is the same for everyone regardless of their speed.

The most unintuitive part of this argument may be the assertion that the speed of light is *always the same* in every frame of reference. If one accepts that statement, the rest is fairly obvious.

One can use algebra and the Pythagorean theorem (which we

just demonstrated through physics) to arrive at time dilation

$$\Delta t = \frac{\Delta \tau}{\sqrt{1 - v^2/c^2}}$$

In other words, the time measured on the ground is longer by a factor of $\frac{1}{\sqrt{1-v^2/c^2}}$.

Statistical Mechanics

Statistical mechanics seeks to describe systems with large numbers of particles, even when the behavior of individual particles cannot be described precisely.

The fundamental quantity in statistical mechanics is the *entropy*,

$$S = \ln \Omega$$

where Ω is the number of possible configurations of the particles. For instance, one can try to count the number of ways there are to arrange a system with a given total energy, E, by counting the number of ways one can distribute the energy among the particles. This number is Ω and the entropy is its logarithm. The fundamental assumption is that *each configuration happens with equal probability.* This seems like the most natural assumption one could make, yet it can lead to surprising predictions.

Puzzle

I have picked a number between 1 and 1000. You can guess a number, and I will tell you if it is equal, above or below the selected number. Try to find the number in the least possible number of guesses.

Solution

You may be familiar with a "binary" search, which involves always trying to guess the middle of the possible range. This is the best approach because you are effectively minimizing the entropy of the system.[27] Basically, you want to put yourself in the position of gaining as much information about the situation as possible from each guess. So dividing the range equally is the optimal strategy since it allows you to eliminate half the possibilities on any one question, which is the best you can do. Any other splitting may land you with the bigger than half of the possibilities left over. So one tries to optimize the worst case scenario.

Puzzle

There are 12 coins. You are told that one of them is counterfeit and is either heavier or lighter than the others, but you are not told which coin it is or whether it is lighter or heavier.

[27]In each step, you divide the set of possible numbers, N, into two parts, $N = N_1 + N_2$, depending on the question you ask. The best question to ask to minimize the leftover possibilities is to minimize the entropy of the possibilities, i.e., to minimize $p_1 log(N_1) + p_2 log(N_2)$ where $p_i = N_i/(N_1 + N_2)$ is the probability we land in a given set (subject to fixing $N_1 + N_2 = N$). That line of reasoning leads to $N_1 = N_2 = N/2$.

You have a balance that will tell you if the two sides are equal, lighter, or heavier. What is the minimum number of weighings that you need to figure out which one is the counterfeit coin and whether it is heavier or lighter? What if there were 500 coins instead of 12?

Solution

Here, again, we rely on the approach of dividing the possibilities into equal portions, just as in the previous puzzle. There are 24 possibilities given that there are 12 coins in total, each of which may be the counterfeit, and the counterfeit can be either heavier or lighter than the rest leading to $12 \times 2 = 24$. And each time we weigh one group of coins against the other, there are three possible outcomes: the first group could be lighter, heavier, or the same weight as the second. So we need to devise a strategy for weighing which will divide the set, as equally as possible, into three groups. The best we can do would be

$$24 \to 8 + 8 + 8$$

$$8 \to 3 + 3 + 2$$

$$3 \to 1 + 1 + 1, \quad or \quad 2 \to 1 + 1 + 0$$

This offers a guide as to how we should proceed. The best we could possibly do would be to identify the counterfeit coin and determine its weight in just three weighings because $3^3 = 27 \geq 24$. To get the first splitting $24 = 8+8+8$ you can easily convince yourself that you can split the 12 coins into three bunches of 4. Then, weigh two of the bunches on the balance. If the first bunch is heavier than the second, you know that either one of the 4 coins in the first group is heavier than average or that one of the 4 coins in the second group is heavier than average, which gives 8 possibilities. A similar outcome holds if the first bunch is lighter. If the first two bunches weigh the same, then you know that the counterfeit lies somewhere in the third bunch which could have the heavier or lighter counterfeit, again 8 possibilities. The subsequent weighings can be performed in a similar fashion, the details of which we leave to the reader.

For 500 coins, the approach is similar, as well. The answer for the least possible guesses is 7. There are 1000 possibilities in this case, just as there were 24 possibilities with the 12 coins. Each weighing of bunches of coins has three possible outcomes, and hence divides the space of possibilities into three parts. Therefore, one needs at least 7 operations to isolate each possibility, because $3^6 < 1000$ and $3^7 > 1000$. You would want to split the possibilities into 3 equal groups (to the extent possible) at each step. There may not be a way of doing this exactly. So we have not proved that the task can be accomplished in just 7 steps, but we have shown that it cannot be done in any fewer.

Puzzle

You have 100 bottles of wine, one of which is poisoned. Your friends are willing to help you figure out the poisoned bottle, but the effects will not be realized until 24 hour later, and you are throwing a party in 25 hours. What is the minimum number of friends that you must recruit in order to determine which bottle is poisoned?

Solution

You need only 7 friends. Label the 100 bottles by binaries. This will require at most 7 digits because $2^7 = 128 > 100$. You have the n-th friend drink only from the bottles where the n-th binary digit of the bottle is 1. From this, you can figure out which bottle had poison by finding its binary by putting a 1 for each digit corresponding to a friend that was poisoned, and a 0 elsewhere. These people, of course, need to be pretty good friends to volunteer for this assignment, and hopefully the poison is not deadly!

Puzzle

Suppose we are playing the following game: we want to throw a ball at the wall so that it bounces up and hits a certain object on the ceiling (ignoring gravity in this case), as in Fig.66. Where do we aim at the wall?

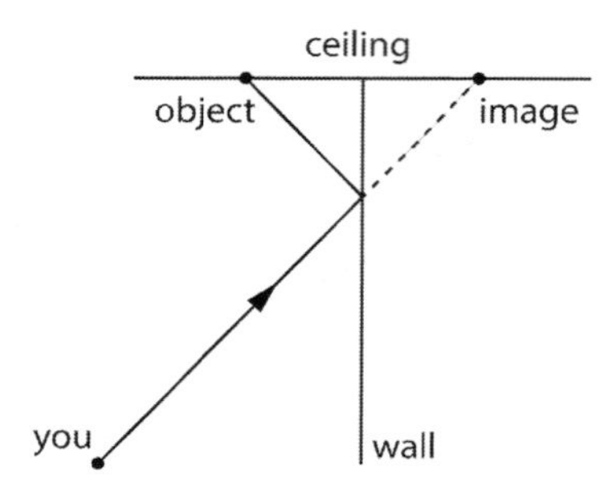

FIGURE 66. You can use a mirror on a wall to correctly aim at an object.

Solution

We can put a large mirror on the wall and aim for the image of the reflection of the target.

Puzzle

A lifeguard standing on a beach needs to rescue a drowning swimmer (Fig.67). Time, in this case, can be a matter of life or death. The lifeguard travels at speed v_1 on land and v_2 in water, and–as is to be expected for a land-dwelling creature–v_1 is greater than v_2. Which path should he follow to minimize the time it takes to reach the swimmer?

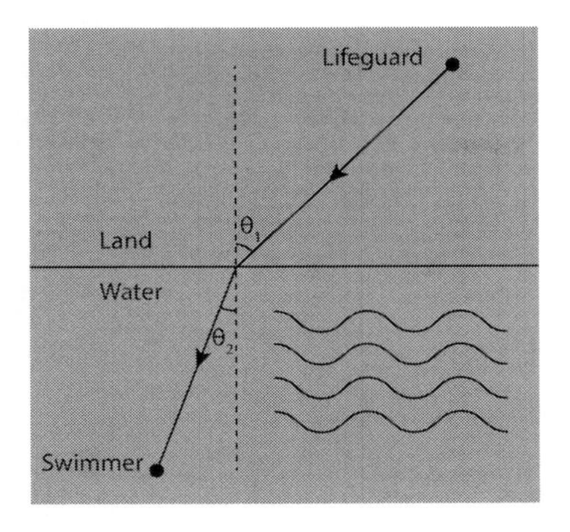

FIGURE 67. Which path does the lifeguard take to get to the drowning swimmer as fast as possible?

Solution

The lifeguard can move more quickly over the ground than in the water, so he needs to determine the right spot from which to start swimming. The problem can be viewed as figuring out the optimal angles he should follow from the beach to the shoreline and from there to the swimmer. The answer to this question can be obtained from Snell's law:

$$\frac{\sin\theta_1}{\sin\theta_2} = \frac{v_1}{v_2}.$$

Snell's law is typically used to describe how light (or any other wave) changes direction, or refracts, as it moves through one medium (with velocity v_1) to another (with velocity is v_2). The law tells us that the path light takes, while traveling through different media, is the one that leads to the shortest total time. But it could just as well apply to the above example of a lifeguard who moves at different speeds on land and water. We could solve this problem by using calculus to minimize $t = \frac{\ell_1}{v_1} + \frac{\ell_2}{v_2}$, where l_1, l_2 denote the length of the paths on the land and in the ocean, respectively.

But we can also devise a mechanical system that enables us to solve the problem without resorting to higher math. Consider a table, with holes cut at two locations that correspond to the positions of the lifeguard and the swimmer. Suspend a mass of $m_1 = \frac{1}{v_1}$ at the lifeguard position and a mass of $m_2 = \frac{1}{v_2}$ at the swimmer position, connected by strings, as in Fig.68. Place a rod in a position that corresponds to the shoreline. A ring that can slide freely along this rod is attached to the strings holding the two masses.

We can now find the solution to our original problem by identifying the configuration it settles to, i.e., the configuration with the lowest potential energy. Since the total lengths of the strings are fixed, to minimize the potential energy, you have to maximize the lengths under the table, which are under tension from the suspended weights. (Potential energy here is represented by mgh, where h is the height of the weights from the floor). Note that the length of each segment of string is fixed. Therefore, the potential energy is minimized when $m_1 g\ell_1 + m_2 g\ell_2$, i.e., when

$\frac{\ell_1}{v_1} + \frac{\ell_2}{v_2}$ is minimized. Therefore, the best path–for the strings and, by analogy, for the lifeguard too–is the configuration at equilibrium. What will this be? Let T_1, T_2 denote the tensions on each string, which are $m_1 g$ and $m_2 g$, are proportional to $1/v_1$ and $1/v_2$ respectively. At equilibrium, the horizontal forces along the rod have to cancel (otherwise the ring will slide on the rod): $T_1 \sin\theta_1 = T_2 \sin\theta_2$. This shows that $\frac{\sin\theta_1}{\sin\theta_2} = \frac{v_1}{v_2}$. We have thus shown that Snell's law minimizes the time–a fact that might be of benefit to the drowning swimmer.

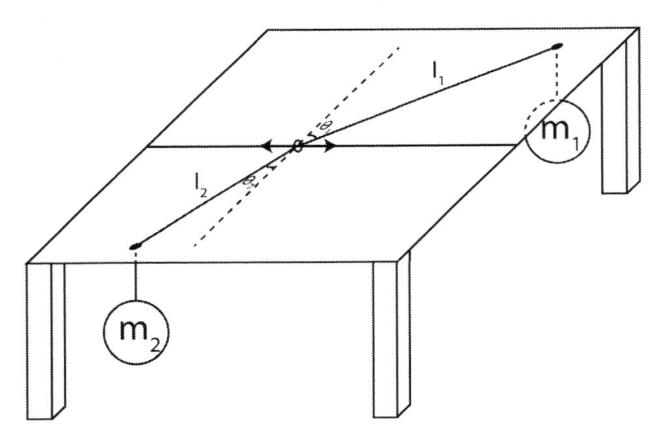

FIGURE 68. A physical model to help the lifeguard find the best path.

The tie-in between string length and light, which is what Snell's law originally addressed, is this: light follows the path that will result in the minimum travel time just as the lifeguard finds the path which leads to shortest travel time.

7. Counter-Intuitive Physics

By now it is probably apparent that not all aspects of physics are intuitive. In fact, some of the most exciting things about physics involve cases where our intuition positively fails us. Sometimes our intuition tells us that a situation cannot possibly be true, and yet it is. Let us start our exploration of counter-intuitive physics by going back to the subject of buoyancy.

Buoyancy Revisited

Many people think that modern physics is unintuitive, and there is certainly some truth to that sentiment. Viewed from outside, the field is getting stranger and stranger. Nevertheless, unintuitive physics is by no means a contemporary phenomenon; it began much, much earlier. Buoyancy, to take one example, has been known for more than 2,000 years, and even after all that time the subject remains largely unintuitive. For how can a very heavy ship be held up simply by the buoyant force of water? Of course, we saw in the last chapter that this concept can be made intuitive if one thinks about it correctly. We might call that a matter of correcting our intuition to make the unintuitive part of physics feel more intuitive. Yet it still seems very strange, especially when contemplating supertankers nearly a half-kilometer long that can carry cargo weighing more than a half-million metric tons!

We all know that a helium balloon will rise due to buoyancy. But what happens if you have a helium balloon in your car and suddenly stop? While you (the driver) will lean forward–hopefully restrained by a seatbelt–the balloon will move *backwards*. What if you turn? It will move inwards, unlike what you may think. This is because, in both cases, as viewed from the perspective of passengers in the car, there is an effective acceleration, and the balloon moves in the direction opposite to that acceleration relative to the rest of the air, which is more dense, as would be expected based on buoyancy.

There is a more extreme version of this, the so-called "Archimedes' Paradox." Suppose you have a huge ship and only a few buckets of water. Can you float the ship using only that much water? The answer is astonishingly yes! Since buoyant forces are local, it suffices to cover the bottom of the ship with just a very thin layer of water, as in Fig.69. Indeed, the argument for the buoyancy force we reviewed in the last chapter did not demand that there be a large reservoir of available water.

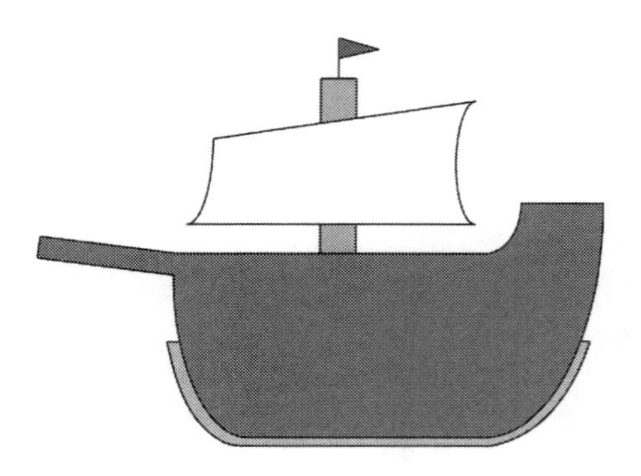

FIGURE 69. Archimedes argued that you could float a ship with a bucket of water if you had a container that fits the shape of the ship.

While there were already many unintuitive things in classical physics, the number of unintuitive things has increased rapidly since we have entered the era of contemporary physics, with much bafflement caused by relativity, quantum mechanics, string theory, etc.

Airplanes

We do not question it too much anymore, but the fact that airplanes can fly is quite astounding. We are talking about a huge and massive metal contraption flying in midair! How is this possible?

The airfoil is designed so that the air moving over the wing moves faster than the air below the wing. Bernoulli's principle, which was formulated in the early 1700s, states that $P+\rho v^2/2$ is constant along the flow, where P is the pressure, v the speed and

ρ the density of the fluid. Therefore, the higher the v, the lower the P. The pressure over the wing is lower than the pressure under the wing, because the wing is designed so that the air moves faster over the wing. This creates a net upwards force, giving the airplane a lift.

The Bernoulli principle, despite its unintuitive consequences, has a simple origin: It is essentially the conservation of energy as applied to laminar flows: Increases in flow velocity (and hence kinetic energy) are caused by the work done due to the net change in pressure and therefore the change in energy is accompanied by decreases in pressure.

Puzzle

You are given a (small) powerful air blower and a (big) light beach ball. How can you suspend (as in Fig.70) the beach ball in midair using only the air blower?

Solution

One may initially think that aiming the blower upwards is the way to go, but that creates an unstable equilibrium, and the ball will quickly fall.

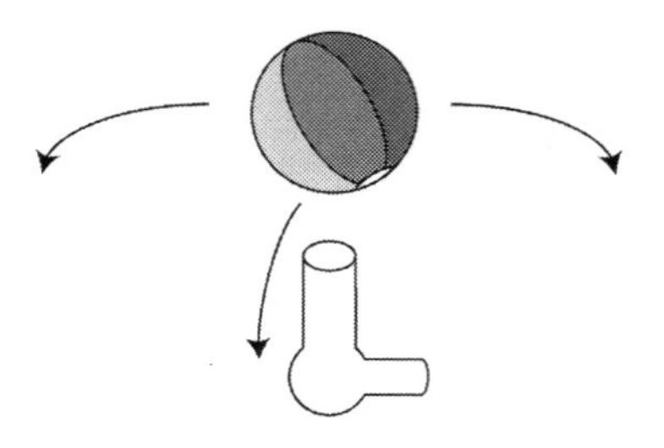

FIGURE 70. Can you suspend the beach ball with an air blower?

Surprisingly, aiming the airstream just above the ball (as in Fig.71), rather than below, will create lift that pushes the ball up. As we have just discussed in the context of airplanes, pressure and velocity are related by the aforementioned Bernoulli principle. So when you increase the velocity of the air above the ball, the pressure decreases and the air pushes the ball upwards countering its weight and leading to a stable situation.

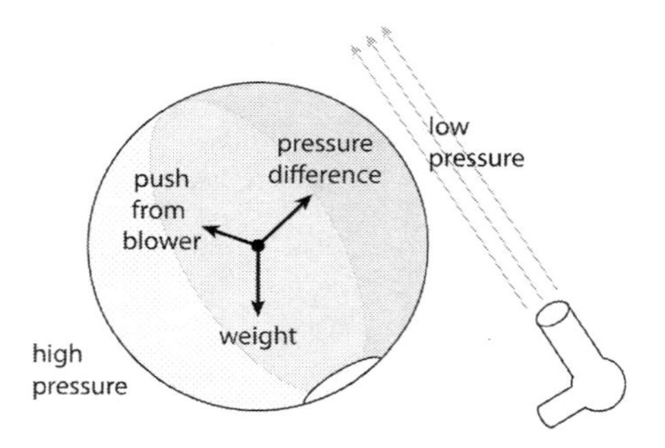

FIGURE 71. To suspend the ball in air, you need to aim the blower just above the ball rather than below it!

Why is the Night Sky Dark?

This is not quite as obvious as it sounds. Back in 1576, Thomas Digges did a simple calculation, which brought to light (almost literally as we shall see!) a famous puzzle in the history of science–the dark night sky paradox. Assuming that the intensity of a light source decays by the inverse square law, and that the universe is relatively homogeneous so that the density of stars is roughly constant everywhere, one can calculate that the intensity of light from stars should be blinding. Under these conditions, in fact, the intensity would be *infinite*, which is contrary to what we know from every day (and "every night") experience: We would not have a dark night sky!

As to why this would be the case under the above assumptions, let us suppose that the number density of stars is ρ. Then the number of stars contained within a spherical shell of radius R and thickness dR is approximately $4\pi R^2 \rho\, dR$. However, the intensity is inversely proportional to R^2. In other words, $I(R) = I_0/R^2$ for each star, so the intensity we receive from a shell of thickness dR a distance R from us is approximately $(I_0/R^2) \cdot (4\pi\rho R^2 dR) = 4\pi I_0 \rho dR$. If we integrate dR from 0 to ∞, we get $\int_0^\infty 4\pi I_0 \rho\, dR = \infty$. In other words, the light from stars should be infinitely bright. This is obviously a problem, as something here is not adding up.

What is the answer to this puzzle? One way out of the

dilemma would be to suppose that our location is special and that stellar density is not constant throughout the universe. Newton offered a different solution–namely that the universe is *finite*. A simple calculation confirms that this conclusion leads to a result that is consistent with our observations. This problem demonstrates, moreover, the value of paradoxes in pointing us towards deeper aspects of physics, including hints regarding the finite nature of the universe.

The modern resolution of this paradox is that the universe is expanding. As a result, light is redshifts as it travels through space, which decreases its energy. We also know that the universe has a finite age, which means that only the light from a finite number of stars would have had time to reach us. These two factors combine to resolve the paradox. In other words, even if the universe is actually infinite in size, the effective size of the universe accessible to our view, is finite. We can see from this exercise that the universe is not infinitely old, which suggests the universe may have had a beginning. So the night sky is dark because the age of the universe is finite!

Maxwell's Equations

As we have already discussed, Maxwell found a way of unifying the theories of electricity and magnetism. His equations led to the conclusion, as we would phrase it now, that there are waves moving through empty space at the speed of light.

Maxwell, however, never thought of it in this way. According to his physical intuition, waves came about by jiggling things, which means there cannot be waves in empty space. For him, the notion of having a solution in empty space did not make sense, so he tried to construct an explanation in terms of a hypothetical medium called the "ether," which turned out to be wrong. Sometimes, our intuition can mislead us in some ways, while putting us on the right track in other ways.

Einstein's Theory of Relativity

Relativity is full of paradoxes, but arguably the most paradoxical one is the law of addition of velocities. According to Newtonian physics, if an object moves with velocity v_1 in one reference frame, then in a different reference frame moving with velocity $-v_2$ with respect to the original, the object should move with velocity $v_1 + v_2$. However, this turns out to be incorrect, especially for objects moving at very high speeds. The correct velocity addition formula according to the special theory of relativity is

$$v = \frac{v_1 + v_2}{1 + v_1 v_2 / c^2}$$

where c is the speed of light. In particular, if $v_1 = c$, we get $\frac{v_2 + c}{1 + v_2/c} = c$ independently of what v_2 is! So, the light has the same speed in every reference frame.

Puzzle

Is it possible, at least in theory, to make a time machine? The idea is to make the design of a spaceship so that you can travel at least in one direction in time. Which direction of time travel is possible, and how do you go about making such a device? Provide the basic specifications of your design if the time travel is to take you on a 1,000-year journey in time as measured on Earth, and you would also like the journey to be just long enough for you to watch a 2-hour movie!

Solution

Traveling back in time violates the principle of causality and is not allowed in any physical theory. However, you *can* travel into the future by going out at a sufficiently high velocity and then coming back. The requisite velocity turns out to be about $(1 - 2.6 \times 10^{-14})\, c$. This follows from the concept of time dilation, which we discussed in the last chapter, where the dilation factor is $1/\sqrt{1 - \frac{v^2}{c^2}}$. So, the design needed to accomplish this task requires a rocket that can travel with such a high speed relative to Earth. Therefore, we can travel along a circle with a circumference of 1,000 light years with this speed, starting and ending on Earth, and it will give us about 2 hours to watch our movie! Along the way, we will have traveled only about 1/50-th of the radius of our galaxy, the Milky Way.

Actually, this is not as impractical as one may think. To

accelerate your mass to that kind of speed, you only need an amount of energy equivalent to your mass. For someone with mass about 100 kg, you can produce that kind of energy from a nuclear energy source. And we can accelerate slowly, without killing whoever is on board. One may wonder why this has not been done yet!

This is the subject of the famous *twin paradox.* According to Einstein's special theory of relativity, all inertial reference frames are equivalent, so how can one twin age differently from the other? The answer turns out to be that one twin must undergo acceleration in order to come back, so that breaks the symmetry between them.

Thinking about this problem can lead us to an additional paradox: What if the universe is *periodic*? For instance, what if it is shaped like a cylinder, and a spaceship can eventually end up at the same position, even by traveling at constant speed without acceleration? In that case, what happens to the twin paradox?

The answer turns out to be that there is a *preferred* frame in which the universe is truly periodic *in space* without shifting time as one goes around the space. And that is the frame in which you would age the fastest.

A Classic Experiment

If you put a tennis ball on top of a basketball and drop them together, the tennis ball will theoretically shoot up roughly nine times higher than the original height from which it was dropped, assuming the collisions are elastic. The fact that the tennis ball bounces so much higher seems completely unintuitive, yet it follows from a simple application of the laws of conservation of energy and momentum. Here's how it works: Let us suppose the basketball and tennis ball are both falling down and about to hit the ground with the same speed, v, and let us further suppose that during their descent, the tennis ball becomes displaced, ever so slightly, from the basketball. The basketball will hit the ground first and–because the collision is elastic–start going up with speed, v, while the tennis ball is still falling with the same speed. Their relative speed at that moment is $2v$. A moment later, the basketball will hit the tennis ball, sending it flying upward. The basketball will continue moving upward with a speed of v because the tennis ball is so light by comparison that its impact is negligible. The tennis ball's collision with the basketball is elastic too, which, as it turns out, means that the two objects will necessarily maintain the same relative speed, $2v$, and to do that the tennis ball must be moving upward at $3v$. Since the maximum height attainable is proportional to the square of the initial speed, the tennis ball will travel about nine times higher than its heavier counterpart.

Paradoxes in Quantum Mechanics

Relativity and quantum mechanics both emerged more than a century ago. While relativity is strange, quantum mechanics is even stranger. One hundred years later, aspects of this field continue to baffle the world's leading physicists, and no relief is in sight!

One of the first paradoxes that triggered the advent of quantum mechanics was the problem of blackbody radiation. If you consider radiation emitted from a box, the classical picture is that at a temperature T, each mode has energy $\frac{1}{2}kT$ where k is the Boltzmann's constant, according to statistical mechanics. But there are infinitely many harmonic modes for radiation waves in a box, so the energy should be infinite. This is similar to the issue of infinite intensity that we discussed in the context of the night sky being dark rather than overwhelmingly bright. Planck resolved this dilemma by suggesting that energy is *quantized* in multiples of $\hbar\omega$, where ω is the frequency of the radiation. He showed that this assumption alone was sufficient to resolve the paradox. Basically what happens is that for frequencies where $\hbar\omega >> kT$ they will not be produced and thus we have effectively a finite number of frequency modes for radiation.

This insight was an important step toward the development of quantum mechanics, which comprises some of the most counter-intuitive portions of our physical laws as we know them today. The unintuitive aspects start with the very postulates of quantum mechanics: particles are like waves, and we cannot ascertain physical phenomena with certainty, but only probabilistically. There is a probability density function (which is the square of the particle's wave function) for determining a particle's position. So the uncertainty in position is not due simply to the inadequacy of our measurement apparatus, but is instead an *inherent* aspect of the particle. In fact, the result of an experiment depends on what you measure: measurement thus becomes an important part of the theory.

Feynman is credited with saying something along these lines: "Anybody who claims to understand quantum mechanics is lying." You can be a top practitioner of quantum mechanics, as Feynman surely was, without having an instinctual feel for the subject. In the same way, we may sometimes use a formalism in physics, often repeatedly so, without completely internalizing the underlying ideas.

The probabilistic nature of quantum mechanics raises some philosophical questions concerning determinism and free will. While those connections are somewhat speculative, quantum mechanics is inherently counter-intuitive. Einstein made his qualms widely known about probabilistic aspects of quantum mechanics, arguing that "God does not play dice with the universe." To which Bohr famously responded: "Stop telling God what to do!"

The Double-Slit Experiment

The double-slit experiment involves shooting particles through a barrier with two slits and seeing where they land. First, only the top slit is opened, and then only the bottom slit is opened. Finally, both slits are opened (Fig.72). You might expect that the distribution of particle paths when both slits are open is the summation of the distributions when each slit is open, but this is *not* the case!

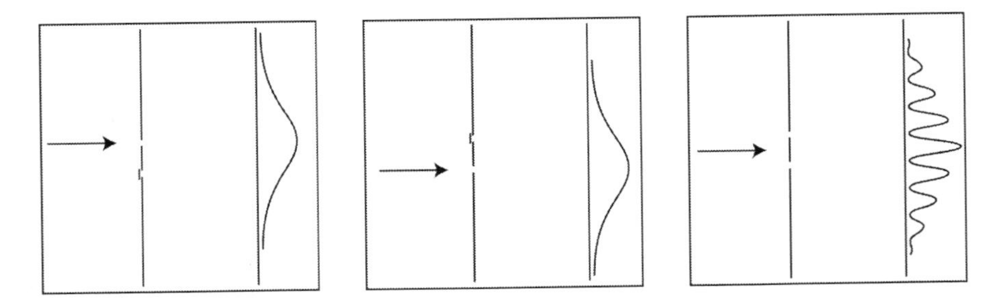

FIGURE 72. Interference pattern of the double slit experiment

When both slits are open, the particle does not go through only one of the slits; it goes through *both* at the same time, as happens with a water wave! In this way, the particle interferes with itself, giving rise to an interference pattern. This occurs even when we send particles one by one: the particles will accumulate in an interference pattern distribution. It is almost as if nature is playing a joke on us. If we try to determine which slit the particle "actually" goes through, say by shining light on the particle and following its course, we will find that it only goes through one of the two slits–but the price we pay is that now the interference pattern disappears! We have interfered with the experiment. The act of measuring, in other words, affects the outcome. We cannot divorce the experimental setup from the outcome, nor can we be strictly passive observers having no influence over events. In a way, this almost sounds like psychology, where the answer we get depends on the kind and order of questions we ask!

There is another example from classical physics, similar in flavor. It is possible to set up the following kind of experiment. If we attempt to shine light through two special glass screens, the light is not observed from behind both screens. However, when you insert a special third screen, in between the two screens, suddenly light is observed!

How is this possible? The screens are *polarizers*. They filter out light in such a way that the electric fields only oscillate in a certain direction. Originally, the two screens have perpendicular polarization axes, as in Fig.73, so the light coming out the first one has no electric field in the direction of the second polarization axis, and, therefore, no light comes out of the second one. By adding the new screen in the middle, and choosing its axis to be at a 45 degree angle relative to the other two screens, as in Fig.74, we project the axis of the electric field to have a different direction. By the time the light reaches the final screen, instead of being perpendicular to it, it has a 45 degree angle relative to the screen and therefore some of it can pass through.[28]

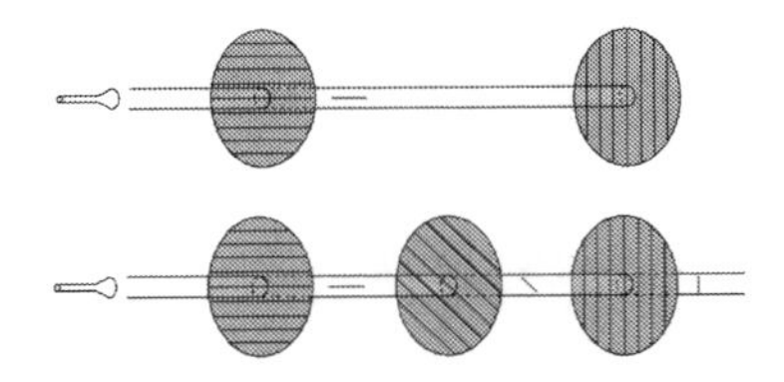

FIGURE 73. The addition of a polarizer in the middle allows some light to pass through.

[28]Interestingly, sailboats utilize a similar principle in order to sail against the wind. The angle of the rudder and the angle of the sail work to shift the orientation of the boat in much the same way that the electric field axes can be shifted using a polarizer. Here, we had a seemingly impossible situation–traveling against a strong wind in a craft that is pushed by the wind–that could be explained rather simply. Quantum mechanics is often like this, too, except that the explanations are rarely that simple.

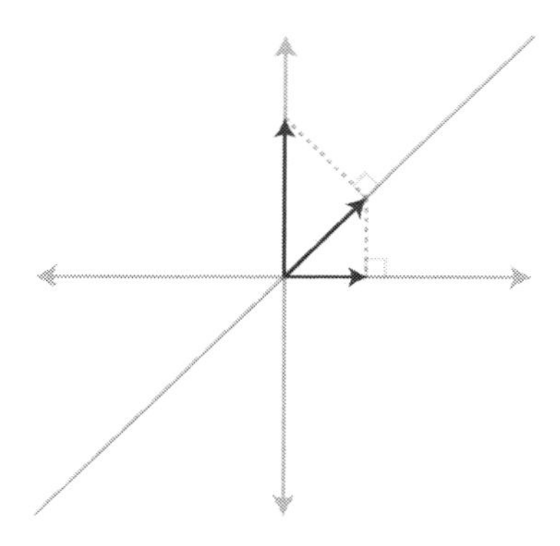

FIGURE 74. The insertion of a middle polarizer leads to gradual rotation of the polarization axis.

What is the connection to quantum theory? Consider an electron. It can have spin up or spin down with respect to any axis. This is already strange; we might think, based on classical reasoning, that there should be "in between" spins. But that is not the case; for each direction you pick, the spin of the electron along that axis is quantized and is either up or down. You can imagine conducting an experiment whereby you ask an electron whether it is spin-up or spin-down. If you ask this question with respect to two different axes, say the x and y, you are forcing spins into those directions. So if the electron's spin is measured to be up with respect to x-axis and then measured with respect to y-axis you find that there is an equal chance for it to be up or down along the y-axis. However, if you measure the spin of an electron with respect to an intermediary axis, $x = y$, before measuring its spin along the y-axis, effectively you force the electron to pick a spin along this intermediate axis which would be more inclined to be in the up direction than down. Now if you measure the spin along the y-axis you find a higher probability for it to be in the up direction, thus changing the outcome of the final measurement. What you see is something analogous to the above example involving light waves in classical physics: when a third screen was added, whose polarization axis was tilted 45 degrees, we projected the electric field in that new direction which changes the outcome at of the last screen.

Indistinguishability in Quantum Mechanics

Another unintuitive aspect of quantum mechanics is the indistinguishability of elementary particles. If I have an electron and you have an electron and we put them in a box with other electrons, neither of us can "tag" our electrons in order to pick them out from the crowd later. It is impossible to distinguish between two electrons. This follows from a symmetry of the joint wave function. If you have an electron in one particular spot and make it trade places with an electron in a different spot, nothing will change in terms of the physics, and that is what we mean by exchange symmetry of identical particles. We could go a bit further and say that nature is in some ways democratic. All the electrons in the universe are identical, and all are indistinguishable; there is no preferential treatment of one electron over another!

The EPR paradox

Einstein famously rejected quantum mechanics, and tried to devise thought experiments refuting it. One of these is the "Einstein-Padolsky-Rosen (EPR) Paradox." Imagine that you have an atom with zero spin, which decays into two particles (e.g., a positron and an electron). See Fig.75. Since the original atom had zero spin, the resulting two particles must have zero net spin. So if the spin of one is up, the spin of the other must be down. But given that we are talking about quantum mechanics, we do not know, a priori, which is up or down until we do an experiment.

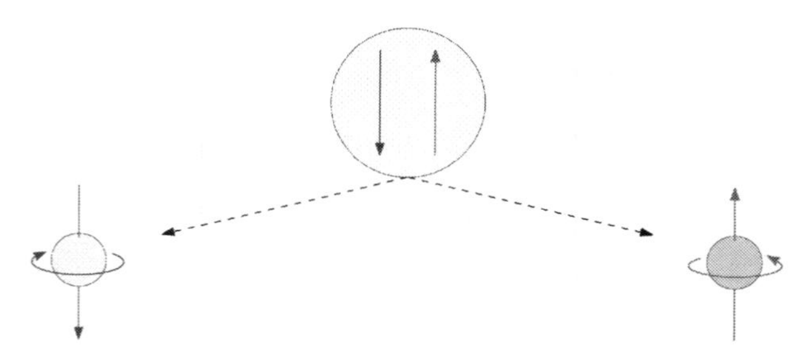

FIGURE 75. The EPR paradox

Now suppose the decay has taken place a long time ago, and the two particles have flown apart, but nobody has measured them yet. Then someone decides to perform an experiment on one of the particles to figure out its spin. She could then predict the outcome of the experiment on the other particle with absolute certainty because the other one will have the opposite spin. In other words, she is determining the outcome of an experiment at a faraway location that another experimenter may be performing at that very time.[29] Einstein called this confounding notion "spooky action at a distance."

Indeed, the concept of non-locality in measurement is an unintuitive aspect of quantum theory. In quantum mechanics, the states and fates of the two particles are "entangled," which means that measuring one is the same as measuring the other, even if the other happens to be very far away–and well beyond the reach of the experimenter. Einstein found this idea wholly unsatisfactory, and he (along with some other physicists) tried to devise alternatives to quantum mechanics where the field's probabilistic features arose not because that was the way nature worked but simply due to the lack of precise information about the state of a system, as is often the case with statistical models too. He assumed there must be some hidden variables, which we did not have precise information about. But if we could come up with

[29]Some have proposed that EPR mechanism could be used to send information faster than light. But this is not possible because no information can transfer faster than the speed of light. The EPR paradox only shows that physics is non-local.

that information, we could then make an exact prediction for the outcome of an experiment rather than just framing a probabilistic statement.

Later, John Bell found a beautiful way to test this theory quantitatively, formulating an experiment whereby quantum mechanics would predict one outcome and the hidden variables theory (no matter what the variables happened to be) would predict a different one. Recent experiments have confirmed the quantum mechanical picture, thereby ruling out hidden variables theories by direct observation. As a result, we have to live with this strange theory that Einstein was never comfortable with.

Despite the success quantum mechanics has had in explaining our world–in ways that often seem to defy intuition–issues concerning measurement theory in quantum mechanics are still unsettled. Some physicists speculate that quantum gravity might bring some clarity to this situation.

Black Holes

Black holes arise from singularities in Einstein's equations in general relativity. One consequence of that theory is that when you pack enough matter into a given volume, you get a black hole. For example, if you managed to squeeze all the mass of our sun into a region with a radius of less than a few kilometers, you would get a black hole. The escape velocity needed to leave the black hole would exceed the speed of light. Once inside, in other words, nothing can escape, not even light. That is why they are called black holes. The outer boundary, which marks the point of no return, is called the horizon of a black hole. Supermassive black holes are believed to exist at the center of many galaxies, including ours. Astronomers have found abundant evidence for the existence of black holes–not by directly seeing them, but by studying matter on the verge of falling into the abyss. Moreover the gravitational waves unleashed during their mergers have been recently measured in the LIGO experiment.

While scientists continue to learn a lot about black holes, these objects are not that well understood from a theoretical standpoint. We have spent the past 30 years trying to answer the question of what happens when something falls into a black hole (Fig.76), and we still do not really know. Einstein's equations tell us that if an object falls into a black hole, it will arrive at singularity in the center–a point of infinite curvature–at some finite time. As to what happens at the singularity, all bets are off.

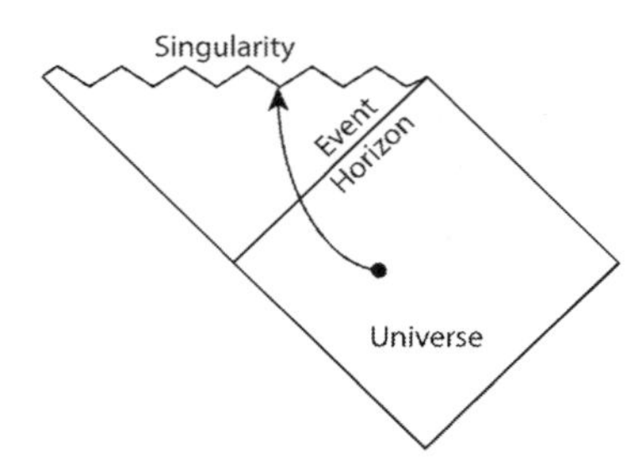

FIGURE 76. It is possible to cross the event horizon and hit the singularity of a black hole in finite time

The fact that solutions to Einstein's equations show that the curvature of space-time becomes infinite in some locations is an indication that this theory is not adequate, on its own, to fully describe black holes. Einstein, himself, refused to believe that black holes could exist even though we are now but certain they exist. Moreover, many physicists now believe that, in order to resolve these curvature singularities that lie at the heart of black holes, we need a broader theory, which integrates both general relativity and quantum mechanics.

This has proved to be a difficult challenge, though considerable progress has been made. Stephen Hawking showed, roughly half a century ago, that black holes can emit Hawking radiation due to quantum effects. Hawking, building on the work of Bekenstein, demonstrated that the surface area of a black hole is related to its entropy (the entropy contained within it), which is, in turn, related to its mass. The information paradox arises because objects falling into a black hole cannot get out. But a black hole radiates energy in a thermal way, as Hawking told us, which means that it will eventually disappear, giving out no information and thus destroying all the information that the objects falling into it had carried. This poses a potentially serious problem, given that quantum mechanics states that information cannot be lost. Our current thinking holds that this information is somehow retrievable, but we do not know enough about the insides of black holes to explain the mechanism that would allow us to retrieve it. This is one of the many reasons that black holes are considered to be among the most enigmatic and unintuitive objects in the universe.

Holography

Holograms are two-dimensional images that convey the illusion of three dimensions. Holography, more broadly, deals with systems that possess one fewer dimension of freedom than they appear to. What does this have to do with black holes? Well, we said that the entropy of a black hole is related to its surface area rather than to its volume. So in this case we appear to be missing one dimension, as if all the information inside the black hole is secretly encoded on its surface (or event horizon)–just like a hologram. See Fig.77. In this way, a three-dimensional problem has suddenly been reduced to a two-dimensional problem. And this line of thinking has become an important source of inspiration to physicists trying to relate gravity in one setting to a physical system of one lower dimension. The principle at play here, known as holography, lies at the heart of some of the most exciting work going on in theoretical physics today. And it is fair to say that much of the impetus behind this work came from thinking about black holes–the impossible objects, which first emerged from calculations in 1915, that now seem more important than could ever have been imagined.

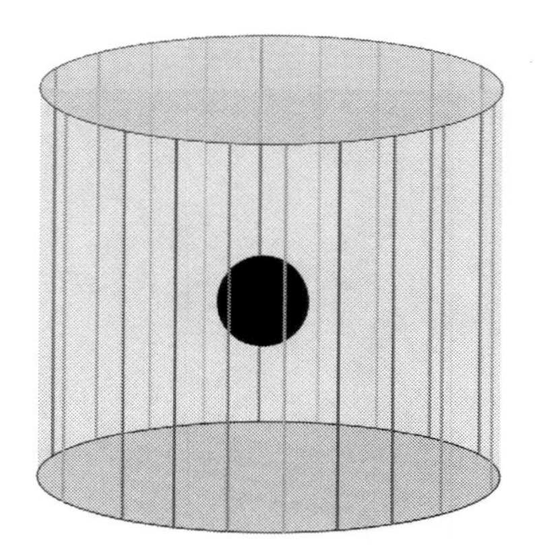

FIGURE 77. Objects inside the cylinder, including black holes, can be described from the perspective of the boundary of the cylinder. This is the concept of holography originally initiated by Gerard 't Hooft and Lenny Susskind.

8. Naturalness in Physics: Dimensional Analysis

A Teachable Moment

A teacher explained the following theorem in class: If M is an $n \times n$ matrix, then $\det(M - \lambda I) = P(\lambda)$ is a polynomial of degree n in λ, called the *characteristic polynomial* of M. It is a fact that $P(M) = 0$, called the *Cayley-Hamilton Theorem.* In other words, a matrix satisfies its own characteristic equation.

A student asked why this statement was true. The teacher responded, "If a matrix doesn't satisfy its own characteristic equation, then whose characteristic equation will it satisfy?"

This anecdote is presented as a joke, albeit one of a rather technical nature. And even though this joke might not make you laugh out loud, it does have some pedagogical value–hinting at the notion of "naturalness," which happens to be the focus of this chapter.

Order of magnitude

What physicists typically mean by the term "natural" is that we expect reasonableness in the laws of physics. Let us start with examples drawn from daily life. Let us ask, for example, how many people have you shaken hands with[30] in the past week? Maybe 5, 10, or 20? Physicists would estimate this number, lying somewhere between 1 or 100, as being on the order of 1,

[30] This question was raised before the COVID-19 pandemic!

which we abbreviate as $O(1)$. It is possible that you somehow shook hands with thousands of people–if you happen to be a politician seeking reelection, for instance, or a celebrity at multiple book signings. Or maybe you shook no hands at all. But this general number, $O(1)$, is most likely in the right ballpark.

This is an example of an *order of magnitude* estimate. Often in physics, we want to estimate a quantity to some reasonable accuracy, but we would like to be able to do that quickly, without going to all the bother of trying to pin the number down exactly. We are going to consider some examples of this in a moment. And for the purposes of our discussion, order of magnitude means somewhere within a factor of 100.

Lots of dimensionless constants come up in physics, numbers like $2, e, \pi, \ldots$. These are all $O(1)$, and we are happy to fit them into our formulas. Whenever you see a formula with some crazy constant, like 10^{25}, you should shake your head and be suspicious. Or at least ask where that number came from? This is, in my view, a happy trend in physics, as it makes life easier for the practitioners while injecting a measure of reasonableness into the field as a whole.

This particular argument, however, does not make any sense for quantities that have units, since we can redefine the scale of the units any way we want. So the expectation is based on quantities which have no dimensions. In other words, we are just saying that dimensionless quantities are expected to be $O(1)$: this is natural. It is really a philosophical principle–as well as, perhaps, an aesthetic one–which cannot be justified by reasoning alone but is reinforced by experience.

Dimensional Analysis

Suppose we are interested in a quantity that is equal to some constant times $A^a B^b C^c$, where A, B, C have independent dimensions and # is dimensionless

$$\text{Quantity} = \# A^a B^b C^c.$$

We can often resort to dimensional analysis to figure out what a, b, c should be. This approach can be particularly powerful when we know that the quantity of interest depends on very few parameters and that there is a unique combination of them that gives us the quantity with the correct dimension. Note that we cannot determine the constant #, but we would guess–and hope–it would be $O(1)$.

Puzzle

Suppose an episode of The Discovery Channel's *MythBusters* tries to duplicate the bus jump stunt in a movie. They make a 1:15 scale model of the bus and bridge. But how should they scale the 60 mph speed of the full-size bus? (Do not use any equations from mechanics; just use the fact that it should depend only on g [the acceleration due to gravity], v [the speed of the bus] and L [the length of the bridge]).

Solution

They should rescale by $\frac{1}{\sqrt{15}}$. Why? Let the critical jump velocity be $v(g, L)$. Then

$$v \propto g^{\alpha} L^{\beta}.$$

The unit of v is $\frac{L}{T}$, the unit for g is $\frac{L}{T^2}$, and the unit for L is L. Therefore, we expect $v^2 \propto gL$, so

$$v \propto \sqrt{gL}$$

Therefore, $v \propto \sqrt{L} = \frac{1}{\sqrt{15}}$

This turns out to be a pretty good formula with $O(1)$ uncertainty.

One can actually calculate the distance L traveled by a particle released with velocity v at an angle θ from vertical. It goes as $v^2 \sim \frac{gL}{\sin 2\theta}$. For moderate values of θ, the constant $\sin 2\theta$ is, once again, $O(1)$.

Radiation From Accelerated Charges

If a charge is accelerated, it radiates light. Let us try to use natural units to help us estimate the power of the radiated light. For starters, power is $\frac{\text{energy}}{\text{time}}$. Power P is a function of charge, acceleration, and the speed of light: $P(q, a, c)$. In natural units, the relevant quantities are:

(1) Force is $F \sim q^2/r^2$ and $E = F \cdot r \sim q^2/r$, so q^2 has units $E \cdot L$.

(2) Power has units $E/T = q^2/(LT)$.

(3) The speed of light has units L/T.

(4) Acceleration has units L/T^2.

Now let us do the dimensional analysis. $T \sim c/a$ and $L \sim c^2/a$. Therefore, we can guess, based on the only combination that yields the correct dimension:

$$P \sim \frac{q^2 a^2}{c^3}$$

What's the right answer? It turns out that we are just missing a factor of $\frac{2}{3}$:

$$P = \frac{2q^2 a^2}{3c^3}$$

This is called Larmor's formula. Again, our estimate was only off from the exact answer by an $O(1)$ factor.

Scaling and Conformal Field Theories

Take a region of a plane and rotate it about an axis to get a three-dimensional object as in Fig.78. If the length along the axis is L, then we would guess that the volume of the resulting object is $\approx L^3$, with some $O(1)$ constant of proportionality. This could be wrong if the shape is pathological or degenerate, but it is generally correct.

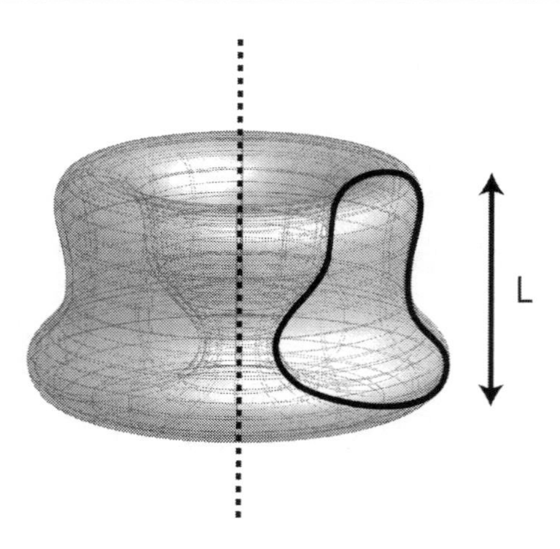

FIGURE 78. Rotating a curve of diameter L around a nearby axis leads to an object with volume of order L^3.

If you have a potato and make it 10 times bigger (in length), how would its weight change? It increases by exactly a factor of 10^3. For a fixed shape, if every parameter is scaled by some constant λ, then the volume is scaled by λ^3. This is a precise, mathematical statement expressing the dimensionality. This brings us to another puzzle.

Puzzle

Suppose we scale up human beings by a factor of 100, without changing the material we are made of. Would there be any problems?

Solution

By the preceding remarks, the weight would increase by a factor of 10^6. However, the strength of bones does not change and so our feet can tolerate more weight proportional to the increase in their cross-sectional areas which only scales quadratically with scale–in this case by 10^4. This means that the bones, under the stress of a factor of $10^6/10^4 = 10^2$ increase in pressure, would almost surely collapse.

So we see that our universe is not scale invariant. Our bodies have a particular size (within some range, of course); atoms have a particular size; and stars have a particular size. This is a fact that we are so accustomed to that it seems rather obvious. However, there are theories in physics, so-called "Conformal Field Theories," which are *invariant* under change of scale. In these theories, the quantities of interest transform in a simple way under scaling of the universe. In conformal field theory, there are no objects with mass, because mass would set a scale, as will be discussed below. Having a theory with no scales makes for what is, in some ways, a much simpler theory.

Fundamental Units

In physics, we inevitably have to deal with quantities that have units. The basic dimension-full quantities in physics involve length (L), time (T), and mass (M). There are a lot of redundant dimension-full quantities like charge and temperature. Charge, for example, can be expressed in terms of Planck's constant, $\hbar$ and c the speed of light, which will be discussing momentarily. Temperature can be expressed in terms of energy through the formula, $E = kT$, where k is the Boltzmann constant and T is the temperature. k turns out to be convenient in thermodynamics, but it does not add a new dimension to physics. Physicists could have continued to talk about temperature in terms of joules, a unit of energy, without ever discussing k, though there's usually nothing wrong with choosing a more expedient approach. At certain junctures, physicists thought they had found a new fundamental quantity, only to discover that it was related to the original three.

So why are there only three independent dimension-full quantities, L, T, and M? I know of no deep explanation of this fact; it just seems to be an intrinsic feature of our universe. Note that you could choose three other independent combinations of L, T, and M to serve as basic quantities, but regardless of how you choose, there will always be three of them.

An amazing result of modern physics is that nature seems to choose her own fundamental units for L, T, and M. It turns out that these three units are related to the three fields of classical mechanics, electromagnetism/relativity, and quantum mechanics. What we mean by that is that each of these subjects introduces a fundamental unit of nature:

(1) Newton introduced the gravitational constant, G.

(2) Electromagnetism and relativity introduced the speed of light, c.

(3) Quantum mechanics introduced Planck's constant, $\hbar$.

It turns out that $G, c, \hbar$ are linearly independent in terms of their dimensions, and so we can write L, T and M in terms of them. Let us see how:

(1) From $F = Ma$ we have for gravitational attraction force $\left[\frac{GM^2}{L^2}\right] = M\frac{L}{T^2}$, which gives us $[G] = \frac{L^3}{MT^2}$.
(2) $[c] = \frac{L}{T}$.
(3) From Planck's equation $E = \hbar\omega$ (with ω being the angular frequency in units of $1/T$), we have $[\hbar] = ET = \frac{ML^2}{T}$.

These are linearly independent because we need to have equal powers of G and $\hbar$ to cancel out the M. That leaves us with $\frac{L^4}{T^3}$, which is independent of $[c]$ in $L - T$ (length-time) space.

Now let us solve for L, T, and M in terms of these new fundamental units.

(1) $L \sim \sqrt{\frac{\hbar G}{c^3}}$. This is called the "Planck length."
(2) $T = \frac{L}{[c]} \sim \sqrt{\frac{\hbar G}{c^5}}$. This is called the "Planck time."
(3) $M = \frac{[\hbar t]}{L^2} \sim \sqrt{\frac{\hbar c}{G}}$. This is called the "Planck mass."

Our preferred unit system in physics is realized by setting all of these to 1. This gives us an elegant, as well as practical, scale of measurement called the natural units or Planck units.

(1) The Planck length is 1.6×10^{-35} m.

(2) The Planck time is 5.4×10^{-44} s.

(3) The Planck mass is 2.2×10^{-8} kg.

The Planck length and time are subatomic, but the Planck mass is huge in comparison, equal in mass to about 10^{19} protons, although it is still small compared to the everyday scales we are used to.

But wait. e, the charge of an electron, also seems fundamental, and we have not used it yet! This means that one of these four constants is not fundamental. It turns out that e has the same units as $\hbar c$ (which the reader is encouraged to check). In terms of the Planck units e is just a number and we may ask whether it is large or small. It turns out that e^2 is $O(1)$ in these natural units:

$$\frac{e^2}{\hbar c} \approx \frac{1}{137}$$

This number is called the *fine structure constant* α. Some physicists have tried to come up with models that expresses this in terms of fundamental mathematical constants like π, e, etc., to no avail. However, it turns out that α is not as fundamental as we think, because quantum electrodynamics tells us electron's charge gets bigger at shorter distances and smaller at longer distances (as we discussed in Chapter 3).

Puzzle

What is the minimum energy of a particle of mass, m, sitting in a box of length L? Classically, it should be zero because the mass is just sitting there. But quantum mechanically, the mass can have some energy due to the fact that it is fluctuating. Use dimensional analysis and the fact that E should only depend on m, L and $\hbar$ to find the order of magnitude of this energy.

Solution

We can construct a unit of energy from the natural units. Recall that $\hbar = [ET] = \frac{ML^2}{T}$. Therefore, $T = \frac{ML^2}{\hbar}$. Since $E = M\frac{L^2}{T^2}$, we get

$$E = \# \frac{\hbar^2}{mL^2}.$$

The constant turns out to be $\# = \frac{\pi^2}{2}$ for the energy of the ground state again an $O(1)$ number. In quantum mechanics, confined particles cannot be absolutely stationary due to Heisenberg's uncertainty principle. With $L \to 0$, the energy increases as the particle gets more confined.

We can see that when Planck's constant goes to zero ($\hbar \to 0$), the minimum energy goes to zero as well. This matches the classical picture. In general, this notion of reproducing classical results from quantum mechanics in this limit is called the *correspondence principle*.

Black Holes

Black holes are among the most enigmatic objects in the universe. But can we compute some basic quantities about them simply by dimensional analysis and naturalness? Solving this puzzle you will see that this is indeed possible.

Puzzle

How much do you have to shrink the sun for it to become a black hole? Use the fact that we know the answer to this question involves gravity, which in turn depends on the mass of the black hole, gravity, and on general relativity (given that it comes from Einstein's theory), which means that it can depend on M, G, and c.

Solution

We are looking for a way to write the radius of the black hole R, which is a length, in terms of G, M, and c, i.e. $R(G, M, c)$. As you can readily check this can be done in only one way:

$$R \sim \frac{GM}{c^2}.$$

After *much* work, which includes solving Einstein's equations, one can get the exact answer:

$$R = 2\frac{GM}{c^2}.$$

This is called the *Schwarzschild radius*. One can compute that if the sun were to collapse into a black hole, the radius would be 2.95 km.

What about the *entropy* of a black hole? Bekenstein argued that it should be proportional to the area of the black hole's horizon. Hawking then computed the precise formula for the entropy.

Puzzle

Estimate the entropy of a black hole of mass M using dimensional analysis.

Solution

Our first step is to make a dimensionless quantity out of area by dividing A by length squared. We want to do this because entropy, being the logarithm of the number of configurations of a system, is itself dimensionless. In other words, we'd like to express the area in terms of Planck units.

$$S_{BH} \propto \frac{A}{\ell_{\text{planck}^2}} \propto \frac{G^2M^2}{c^4}\frac{c^3}{\hbar G} = \frac{GM^2}{\hbar c}.$$

Hawking computed that the constant of proportionality should be 4π (making it exactly a quarter of A in Planck units):

$$S_{BH} = 4\pi \frac{GM^2}{\hbar c}.$$

Symmetry and Naturalness

We said before that physicists are often uncomfortable with numbers that are not $O(1)$. However, this is not always the case. Physicists will accept an exceedingly small number if it is very close to a symmetric point and if the symmetry of the physical system would get enhanced if that number were to vanish (i.e., go to zero). We might consider such a very small number natural because it's so near to the symmetric point. In such cases we would say that number is "protected" by the symmetry, and we would treat it as if it is literally at the symmetric point.

This point can be illustrated by the following example. Assume that Earth is perfectly spherical and that it enjoys exact rotation symmetry. How far away is the center of mass from the center of the sphere? In this situation, clearly that distance is zero due to symmetry.

Now suppose that one person is standing on the surface of the Earth, thereby upsetting the presumed perfect spherical symmetry. Let Δ be the difference between the center of mass and the center of the sphere, and the Earth's radius, R. The center of mass now shifts by a tiny amount proportional to the ratio of the mass of the person divided by the mass of the Earth, which is ≈ 0. So the difference between the center of mass and the center of the sphere is a microscopic quantity–so small as to be essentially negligible–because it is protected by this spherical symmetry. In such a case the fact that Δ/R is so small and not $O(1)$ will not surprise us.

As an aside, some people have actually measured the difference between the center of mass of the Earth and the "center of the Earth" (though it's a bit tricky to define what the latter should mean precisely). According to them, this difference is around 10 cm, which corresponds to a fractional difference $\Delta/R \approx 10^{-8}$. This is another example of a quantity being small due to the fact that it is protected by an approximate symmetry.

We will explore ideas like this in the next chapter. Physicists often ask questions, for instance, about why the mass of the proton is so small, some 10^{19} times less than the Planck mass. This is another unreasonably small number, and we would like to explain it somehow.

There are many candidate "fundamental units," such as the size of the proton or the length or age of the universe. Why aren't they taken as the fundamental units for nature? And what is the relation between these numbers, if any? Dirac explored such questions earlier in the 20th century, trying to understand where these big numbers came from and the ways in which they might be interconnected. Physicists have come up with some new ideas since then, but nobody believes we have the complete answer yet.

This brings out an interesting point about naturalness. It is an important guiding principle, and many theoretical physicists nowadays spend their time thinking about naturality and its role in our universe, while trying at the same time to explain seemingly unnatural phenomena.

9. Unnaturalness and Large Numbers

In the last chapter, we explored the power of naturalness. Dimensionless numbers that appear in physics, we said, are expected to be of order 1 numbers. That makes everything nicer and neater–and generally more tasteful. However, this prejudice does, at times, run contrary to the facts. There are some very big, as well as some very small, dimensionless numbers that crop up in physics. That issue, as we will see in this chapter, turns out to be a rather important aspect of modern physics. For explaining the presence, and persistence, of very large and very small numbers in nature constitutes one of the biggest challenges facing physicists today.

Unnatural Numbers

There are a few dimensionless quantities in physics. For instance, if e is the charge of the electron, $\hbar$ is the Planck constant, and c the speed of light, then as we have already noted there is a dimensionless combination

$$\frac{e^2}{\hbar c} \approx \frac{1}{137}.$$

Unfortunately, this is one of the few natural dimensionless constants in physics that is of a reasonable order of magnitude. A primary motivation in modern theoretical physics is to understand the "unnaturalness" of other constants. For example, the

Planck mass is 10^{19} GeV. In other words, the mass of the proton in the natural units of the universe (i.e., the Planck units) is $m_p/M_{planck} = 10^{-19}$. This is a *tiny* dimensionless number which is fundamental to physics and certainly not order 1!

Interestingly, the natural mass scales fall into roughly three groups, separated by about 30 orders of magnitude altogether, i.e., factors of 10^{30}. In Planck units in logarithmic scale we have

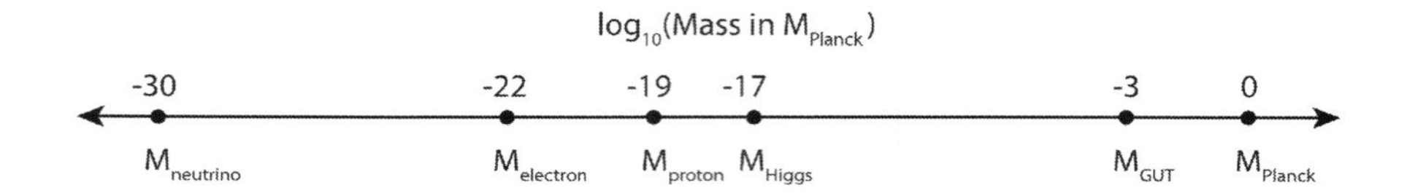

The scales plotted here give the mass ranges of elementary particles, which are bunched in the middle of the plot, except for the neutrinos whose mass is very small. In addition, there is another scale, M_{GUT}, which is close to the Planck scale that we discussed before. The charges (like the electron's charge e) are said to "run" with the energy scale, and this running of the coupling–the dependence of charge on energy or distance–is logarithmic. If we plot the dimensionless constant $\frac{e^2}{\hbar c}$ (and its weak force and strong force equivalents) as a function of energy, as we have discussed before, we see that they become equal at some high energy. Theoretical ideas suggest that in that energy regime, these forces become indistinguishable. We say that these forces have been *unified*. The energy scale where this happens is called the grand unification scale M_{GUT}. This energy scale is close to the Planck scale so it is of order 1, which means that no unnatural fine tuning is needed to explain it.

However, the other energy scales are very different from the Planck scale, and this disparity in energy scales does beg for an explanation. Ideas in quantum field theory have led to the prediction that mass scales can be related to one another in an exponential way–a manifestation of the running we mentioned above. In particular, an explanation of the smallness of the mass of the proton compared to the Planck mass is that

$$M_p = M_{\text{GUT}} \exp(-\#\frac{\hbar c}{g^2})$$

where g is the charge for the strong forces. In other words, since the exponent can be a big number but still of order 1, it can lead to a huge hierarchy of mass scales. Similarly, ideas from quantum field theory naturally explain why the masses of electron and similar particles are not too far from that of the Higgs particle, as we discussed earlier. Theoretical ideas also predict that the mass of the Higgs boson is a geometric mean between the neutrino mass and the GUT mass scale.

$$M_H^2 \approx M_\nu \cdot M_{\text{GUT}}$$

The fact the $M_H << M_{Planck}$ constitutes one of the great mysteries of contemporary physics, which is called the "hierarchy problem." As the Higgs particle moves through the vacuum of space, it would interact with many other particles–some spontaneously popping into and out of existence–and each of these interactions could, in principle, make a quantum mechanical contribution to the Higgs mass that would drive it towards the Planck mass instead of being so small. Theorists have proposed a possible way out of this dilemma by postulating an extra symmetry, supersymmetry, that would cancel out additions to the Higgs mass due to quantum effects. This idea proposes that every known particle of nature has an as-of-yet-unseen supersymmetric partner whose mass contributions would almost perfectly offset each other.

However, given that supersymmetry remains an unproven idea, the hierarchy problem still looms large, again asking why

$$M_{\text{Higgs}} \ll M_{\text{Planck}}.$$

We need to explain this disparity–why the Higgs mass is 17 orders of magnitude less than the Planck scale. As we discussed earlier, the Higgs field imparts mass to particles because of a broken symmetry. So one may think that herein lies the answer: $\langle Higgs \rangle$ is of small value owing to symmetry reasons.

This sounds like a nice solution, but quantum theory suggests otherwise. It turns out that when one computes quantum corrections, even if we start with a small amount away from the symmetric value, the quantum fluctuations are still so large that they inexorably push the energy scale back to the Planck scale. Physicists have tried various approaches for taming the quantum fluctuations, so far to no avail. Supersymmetry is the most popular of the potential solutions offered so far, but it has not yet been seen, either directly or indirectly, in the LHC experiments at CERN. So we do not, at present, have a convincing explanation for the hierarchy of mass scales set by the Higgs field and Planck scale.

Archimedes' Cattle Problem

This is a problem posed by Archimedes to those who thought they had mastered math. Archimedes, himself, could not solve the problem. There are two types of cattle, cows and bulls, adorned in four colors: white, black, dappled, and yellow. There are W, B, D, Y cows and W', B', D', Y' bulls, respectively. Their counts satisfy the equations

$$W = \left(\frac{1}{2} + \frac{1}{3}\right) B + Y \qquad W' = \left(\frac{1}{3} + \frac{1}{4}\right)(B + B')$$

$$B = \left(\frac{1}{4} + \frac{1}{5}\right) D + Y \qquad B' = \left(\frac{1}{4} + \frac{1}{5}\right)(D + D')$$

$$D = \left(\frac{1}{6} + \frac{1}{7}\right) W + Y \qquad D' = \left(\frac{1}{6} + \frac{1}{5}\right)(Y + Y')$$

$$Y' = \left(\frac{1}{6} + \frac{1}{7}\right)(W + W')$$

The two additional constraints are that $W + B = k^2$, a perfect square, and $D + Y = \frac{n(n+1)}{2}$, a triangular number. The question he posed is what are the smallest values of $W, B, D, Y, W', B', D', Y'$ which satisfy these equations? One would naively expect the smallest number not to be too big, because the numbers appearing in the above equation are of $O(1)$.

The smallest possible solution for the number of cattle is $\sim 10^{206545}$, which is huge![31] The point is that simple equations, combined with constraints that limit the number of cows to non-negative integers, can naturally lead to astronomical numbers. How can such a simple looking problem have such an un-simple looking answer? Maybe something similar is happening in physics, where natural constraints involving integers are somehow forcing the numbers to get very large. And perhaps the problems of hierarchy that have come to the fore in contemporary physics require a deeper understanding of number theory and its role in physics.

Emergence of Heliocentric Model and Unnaturalness

The idea that Earth is not stationary may appear obvious from the modern perspective, but actually, a lot of evidence pointed towards Earth being stationary. In particular, all the heavenly objects except for a few of them, namely the sun, moon,

[31] see "Das Problema bovinum des Archimedes" by A. Amthor and B. Krumbiegel published in *Z. Math. Phys.* Volume 25 in 1880

and a few "wandering stars" (i.e., planets), were stationary with respect to Earth[32].

When in the 3rd century B.C. Aristarchus of Samos proposed that the Earth is not stationary and revolves around the sun, he was criticized: His model would suggest that not only the Earth would revolve around the sun, but also all the other stars which seemed stationary with respect to the Earth would all have to be revolving around the sun as well! How else could it be, if they appear to be static relative to the Earth and the Earth is revolving around the sun?! And to assume that all the heavenly objects are revolving around the sun sounded like a bizarre situation. However, Aristarchus argued that if these stars are much farther away from us compared to the moon, sun, and "wandering stars" then they will also appear stationary with respect to the sun as well! So a heliocentric model would also look as simple for the stars. But his theory had *an unnatural aspect* to it: How is it that a few of the heavenly objects are so close to us and the rest are almost infinitely far away to avoid a detection of motion as the Earth goes around the sun? This postulate that the ratio of the distance from other stars to us compared to the distance scales of the solar system is such a large

[32]Of course they knew Earth was spinning about its own axis as they could deduce by noting that all the stars appear to be revolving around the Northern Star at night. Almost all the heavenly objects appeared stationary once you took this spinning into account.

number appeared as a problem for this model. This hierarchy of scales is now explained in terms of the structure formation of the stars and the planets.

Number Theory

Number theory is a natural place to come across large numbers. Fermat's Last Theorem (conjecture) states that there are no positive solutions to $a^n + b^n = c^n$ for $n > 2$ which is now proven to be true[33]. Leonhard Euler extended the conjecture in the late 1700s, proposing, among other examples, that there were no solutions to $a^4 + b^4 + c^4 = d^4$ in positive integers. This was *disproved* by Noam Elkies in 1988. The smallest counterexample is:

$$(95800)^4 + (217519)^4 + (414560)^4 = (422481)^4.$$

Is this natural or unnatural? The question, itself, certainly seems natural and involving small numbers. Nevertheless Elkies gave theoretical arguments for why the smallest counterexample is so large (involving 20-digit numbers on the two sides of the equation above) and seemingly unnatural. One may speculate that the unnatural numbers in physics arise from natural questions in number theory.

[33]A proof of the theorem was published by Andrew Wiles in 1995, about 350 years after it was set forth by Fermat.

Card Trick

We have an ordinary deck. The face cards count as 1. Pick a random number from 1 to 10, say n_0. Then lay out n_0 cards from the deck. The number on card n_0 becomes n_1, and we then lay out n_1 cards. The last card becomes n_2. Then lay down n_2 more cards, and continue this process until the deck is done. The object of this trick is to identify the last card we see of the n_f before the deck runs out.

If you try this out with different initial numbers n_0, you will see that all numbers will converge to the same answer n_f! If you did not tell the choice you made, the performer can still pick out the last n_f card that you will get. This gives the illusion that the performer somehow luckily guesses the answer, when in fact all she does is to choose her own initial choice for n_1. This coincidence comes from the fact that once two people converge at some point in the drawing to the same n_i card, they will agree on what the final card is. And the chances are high for the two series of choices to converge at some point. The fact that face cards count as 1 also helps! This game, as set up, is doomed to converge. And so long as there is convergence somewhere along the way, the answer will be the same. The result, however, would seem unnatural if one does not grasp the underlying mechanism.

Could the seemingly unnatural and bizarre numbers we encounter in physics arise due to processes like the one here?

Composition of the Universe

One more constant needs to be mentioned: the cosmological constant, Λ, which has units of $(mass)^4$. As we will see later, it turns out that $\Lambda^{1/4}$ has roughly the same scale as M_{neutrino}; again a quantity that is very small and needs to be explained. We will discuss that later in this chapter. Λ is related to what is called the "dark energy," which is the energy of empty space and is related to the cosmological constant Λ. This is because dark energy leads to an accelerated expansion of the cosmos. It was an amazingly important discovery in physics a couple of decades ago that the composition of energy in the universe is dominated by this mysterious form of energy by measuring this accelerated expansion of the universe. Figuring out where this dark energy comes from constitutes one of the biggest puzzles in all of physics today. The composition of energy in the universe today is as follows:

Energy in	the Universe
5%	matter
25%	dark matter
70%	dark energy

The 5% matter is all the stuff we are made out of. We cannot see the other 95% directly because it does not interact with light, which means that most of the universe is invisible to us. In addition to dark energy, we know dark matter must exist because of its gravitational effects, which have been observed. It is called dark because light does not interact with it much, and it is fundamentally different from the stuff we are made of.

The Geometry of Space-time

Einstein theorized that the geometry of the universe (specifically space-time) should not be taken as fixed and rigid but rather as something that can change depending on the way matter is spread around. The geometry of space-time is determined by the metric, $g_{\mu\nu}$, which provides a way of measuring distance within that space. The distribution of matter, in turn, influences the metric and changes it. At points near large concentrations of mass, the metric will have higher curvature and space will be more curved.

Particles follow *geodesics*, the shortest path between two points—even in curved space. A key contribution of Einstein's general theory of relativity was the introduction of his field equation[34]

$$G_{\mu\nu} + \Lambda g_{\mu\nu} \sim T_{\mu\nu},$$

[34]One of the first tests of this theory was conducted in May 1919. A team led by British astronomer Arthur Stanley Eddington confirmed Einstein's prediction of gravitational deflection of starlight by the sun while overseeing the work of two teams that were charged with photographing a solar eclipse with dual expeditions in Sobral of northern Brazil and Príncipe, a West African island.

where $G_{\mu\nu}$ is the Einstein tensor (which basically describes the curvature of space-time), $T_{\mu\nu}$ is the energy-momentum density, $g_{\mu\nu}$ is the aforementioned metric, and Λ is the "cosmological constant." Originally, this factor (Λ) was not in the equation, and its absence implied that the universe was expanding or contracting. So Einstein put in this factor by hand to find a solution of his equations in which the universe is stationary, with value $\Lambda = 4\pi\rho$, because he assumed (without, as it turns out, correct empirical evidence) that the universe must be stationary. He also chose the space to be curved–perfectly balanced against expansion or contraction by the cosmological constant.

In the Planck units, the Λ that Einstein found is an extraordinarily small number, on the order of about 10^{-120}, but Einstein chose it anyway to get a static universe. Interestingly, this solution to keeping the universe static was rather unstable: If Λ were just a tiny bit smaller or larger, Einstein's solution would either expand or contract. A Jesuit priest, Georges Lemaître , who was also a mathematical physicist, advanced a different model in the 1920s in which the universe started from a primordial atom and expanded. Einstein rejected that theory! A bit later in the 1920s and 30s, Friedmann, Robertson and Walker made a precise model for the expanding universe. Facts concerning the universe's first fraction of a second are still being debated today, but we believe that the general picture advanced by Lemaître and Friedmann, Robertson and Walker–known as the Big Bang theory–is correct.

Later, the amateur astronomer Hubble measured the expansion of the universe by observing that light signals from faraway galaxies are redshifted, which is consistent with an expanding universe. So Einstein then went back and removed the cosmological constant from his equation because that would have naturally led to the prediction that the universe is expanding. He called the inclusion of the cosmological constant in his original equation the "biggest blunder of my scientific life." Einstein could have predicted the expansion of the universe had he placed more faith in the original form of his own equations.

Years later, even to the late 1980s, physicists were still asking themselves: Why is $\Lambda = 0$? The problem is that the quantum fluctuations, according to their calculations, would force a huge correction on the order of magnitude of $\Lambda \sim M^4_{\text{planck}}$. In Planck units, in other words, $\Lambda = O(1)$ and not 0.

Physicists spent many years trying to explain why $\Lambda = 0$, which was the commonly accepted value, despite the fact that quantum fluctuations would seemingly lead to a different answer. The most popular explanation was supersymmetry, which had offered a potential solution to the hierarchy problem, but people still could not quite bend the theory to show that $\Lambda = 0$. Physicists tried to use supersymmetry, supergravity, and string theory, but nothing helped to reduce Λ to exactly 0.

Fast-forward to the late 1980s. I was at Harvard, attending lectures by Steven Weinberg on the cosmological constant. He recalled that some physicists had argued for the value of $\Lambda = 0$ by using the "anthropic principle." Their contention was that our very existence is incompatible with $\Lambda = O(1)$ in Planck units because the universe would only have had a lifetime on the order of Planck time, 10^{-43}s, and we would not have been around to ask the question. Weinberg pointed out that critics of that view regarded the argument as unscientific because it is a postdiction, coming after the fact, and therefore cannot be used to predict anything. However, Weinberg argued that, if interpreted correctly, the approach could lead to a scientific prediction: The anthropic principle, he said, can work only if there are many possible universes, each having a different value of Λ. The universes with big, medium and small values of Λ could not support life, whereas the ones with extremely small or zero Λ could. Now we can use a conditional probability: The Λ of interest to us must be drawn from those that can support life. Given our existence as a precondition, what is the most likely value of our cosmological constant?

The idea is that the value of Λ should be no more fine-tuned than necessary to support our existence. This means that Λ does not need to be exactly 0, but it needs to be of a generic value that would be appropriate to support life. Based on this, Weinberg estimated that $\Lambda \sim \#\rho \sim 10^{-120} M_{\text{pl}}^4$, and he said that this should be observable very soon as it is close to the experimental bounds at the time he was giving his lecture. Less than a decade later, the cosmological constant–with values not too far from what he had predicted based on the anthropic principle–was discovered from astronomical observations of distant supernova explosions! (Ironically, the measured value turned out to be close to Einstein's original prediction for it, based on the false assumption that the universe is static! Removing that constant was his second blunder because unbeknownst to him that term is there with the value close to what he had originally put in his equation!)[35] There are still many physicists who are unhappy with the anthropic principle. It deviates from other laws in physics, but it does follow the scientific methodology to some extent.

[35]In order to use the anthropic argument, Weinberg needed to assume that there were many possible universes. String theorists have developed theories supporting multiple universe solutions, which are consistent with this principle. Each one of these solutions represents a possible universe. This, of course, creates its own headaches because we have not yet found the principle that would enable us to pick the solution representing the universe we live in.

Other Questions

Why is it that the cosmological constant and the matter density today are of the same order $\Lambda \sim \#\rho$? This is a strange coincidence, given that $\Lambda \sim \#\rho$, ρ changes with time as the universe expands, but Λ, being a constant of nature, should not change. That means that the rough equivalence between Λ and ρ is only true for the time we live in now–a confounding situation known as the "coincidence problem." It would suggest that we are now occupying a special time in cosmic history–a proposition that many physicists find unsettling.

It is also curious as to why $\Lambda \sim M^4_{\text{neutrino}} \sim \rho$ today. Clearly these are difficult questions, potentially related to the hierarchy problem, which we have not quite understood. Many theoretical physicists are busy writing models in the hopes of explaining these coincidences.

Distance Scales

So far we have focused on mass scales, but it is also worth talking about distance scales. The smallest is the Planck scale, then there is the proton scale, the radius of the sun, and finally the biggest scale is the radius of the observed universe.

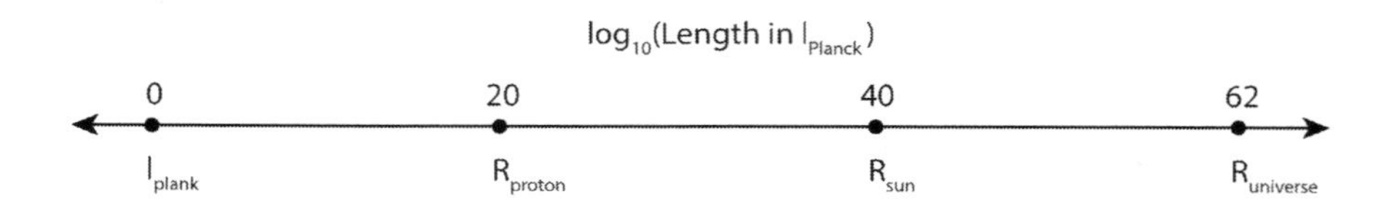

By the way, we should note that mass and length are related. For every mass m we get a length scale R according to $R = \frac{\hbar}{mc}$. In Planck units (where $\hbar = c$) we have $R = 1/m$. So in particular the fact that the radius of proton R_p is about 10^{20} times bigger than the Planck length is related to the fact that the mass of the proton is 10^{-20} times the Planck mass.

Also, it turns out that one can explain why $R_{sun} \sim \frac{1}{m_p m_e} \sim R_p^2 \sim 10^{40}$ in Planck units from first principles. The fact that the masses of a proton and an electron are not too far from one another explains why the distance scale associated with radius of the sun is R_p^2. However, the fact that the radius of the universe *now* is $\sim R_p^3$ has no clear explanation.

Time Scales

There is also a time scale associated with natural units that is related to the ones already discussed. The smallest scale is the Planck time, which is 10^{-43} sec, and the largest observed one is the age of the universe, which is about 10^{62} in Planck units. (Note that this is consistent with the fact that the speed of light is 1 in natural units, and the size of the universe is 10^{62} because distance, of course, equals speed multiplied by time).

One might ask if there is a largest time scale. We do not know. However, based on models from string theory, we have seen no example of a permanently stable universe if supersymmetry is spontaneously broken. Based on the experiments done to date, we would have to conclude that supersymmetry is not realized and is, at best, spontaneously broken. Therefore, to the best of our knowledge the universe will decay. That means, unfortunately, that there is a maximum time for our current universe. We do not know how long this will be, but we expect that our whole unstable universe will eventually decay to a more stable one: Perhaps a bubble of material from the new universe will start expanding and moving with the speed of light. In time, it will overtake and transform the entire observable universe.

I am sorry to report to those who like the status quo that this story does not appear to have a happier ending. Though some readers, perhaps, may find solace in the fact that our current universe might have emerged from just such a transition in the past. And there may well be more transitions in the future.

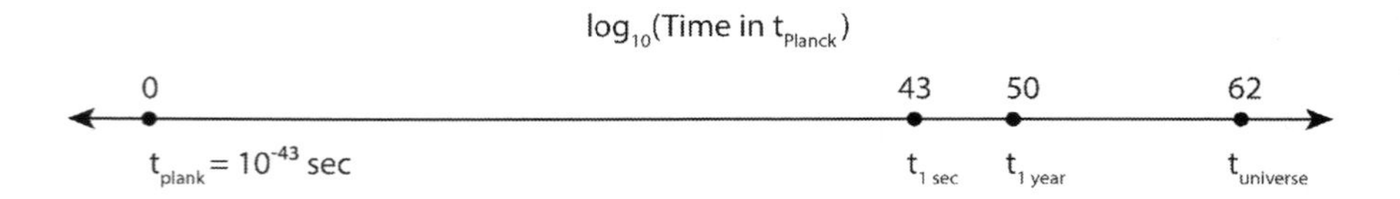

In fact, the cosmological constant gives a theoretical time scale of about $1/\sqrt{\Lambda} \sim 100$ billion years, and it is conceivable (and there are recent theoretical arguments to suggest) that this may set an upper bound to the lifetime of our universe. The fact that the universe is about 14 billion years old now would suggest that we may be at the teenager stage of the cosmic life cycle! If the universe acts responsibly, in other words, it could still have a number of good years–and perhaps even golden years–to look forward to!

10. **Religion and Science**

The interplay between science and religion has a long history, and it would be understating things to say that the relationship has been strained at times. In 1600, for example, the Italian philosopher, cosmologist, and former Catholic priest Giordano Bruno was burned at the stake for espousing heretical views, including the notion that the universe is infinite and contains an infinite number of worlds. In 1633, Galileo Galilei was condemned by the Roman Catholic Church for his blasphemous claims that the Earth moved around the sun. Galileo, who was 69 when this judgment was passed, might have met a fate similar to Bruno had he not renounced his scientific findings. In lieu of torture, imprisonment, and possibly even execution, Galileo was placed under house arrest for eight years until his death in 1642.

It is clear that interactions between science and religion have not always been harmonious, and signs of tension and conflict are still manifest today. Nevertheless, one might say that religion was the first form of science in its effort to make sense of the world in which we live and to explain why things are the way they are. Most, if not all, religions attempt to say something about the physical universe. Furthermore, both science and religion are based on observation, even though the "methodologies"–to the extent that word applies–are completely different. What happens when the views clash? My aim in this discussion is not to come up with any conclusive statement about the relation between the two but rather to review what a sample of scientists have thought about this relation in the past without offering my personal views on the subject. And, as always, I will try to frame the discussion in the context of puzzles!

Basic Questions

There are some basic mathematical and logical puzzles associated with the usual popular descriptions of God. For instance, it is often said that God can do or create anything.

If that's the case, can God create a stone that God cannot lift? Some people who do not believe in God have tried to use this kind of argument to rule out God's existence. This, however, is simply a logic game. It is similar to having a piece of paper with inconsistent statements inscribed on each side. "The sentence on the other side of this paper is false" appears on one side. "The sentence on the other side is true" appears on the other side. With this opposite circularity, there is no possibility for assigning a truth or falsehood value to either sentence. In other words, *it is not true that every statement is either true or false.* This could also be the resolution to the above logic puzzle raised in the context of the existence of God. This reminds me of the following puzzle.

Puzzle

There are two prisoners in two separate cells. Each is given a coin to toss, and each one has to predict whether the other one got heads or tails. If exactly one of them gets it right, they are both freed. Otherwise, they will both continue staying in prison. They can briefly discuss a strategy before they are brought to their separate cells to toss the coins. Is there a "winning" strategy here–one that gets them freed?

Solution

At first sight, it might appear that their fates depend solely on luck. However, there indeed exists a simple strategy that can ensure their freedom. One of the two should predict that the other person gets the same coin toss as he does. The other person should predict the opposite of whatever coin toss he gets. It is easy to confirm that this is a winning approach, even though it might not have seemed possible.

Methodology

You might suppose that mathematics and religion are polar opposites with nothing in common, but that is not entirely the case. Mathematics and religion start with some axioms and definitions that cannot be wholly justified and they proceed from there. Even though mathematics is built upon logic, mathematicians have to start somewhere, and they cannot prove–nor do they try to prove–all of the original axioms upon which their discipline is based. In fact, not only can the axioms not be justified, they cannot even be completed, due to Gödel's incompleteness theorem!

In science, we are not concerned with finding the absolute, unassailable truth but only in establishing the best truth we can at the time–knowing full well that all findings are subject to revision and refinement. Even today, after great advances in science have been achieved and after firm principles of physics have been spelled out, I still do not believe we can issue an absolute statement of truth. In particular, it is hard for me to believe that *any* of the currently accepted principles of physics are exactly correct. Many of them appear to be on the right track, of course, but that is not the same as saying they are exactly right.

Science vs. Religion

From the perspective of many scientists, religion does not enter into the discussion about the nature of the real, observable world and is instead more appropriate for the exploration of moral and spiritual issues. They point out that science is based on testable observations that can be *proven*, whereas religion, is based on beliefs that cannot be fully vetted! However, we can offer some counterpoints to that perspective for science cannot truly disprove religious claims. For instance, a religion is sometimes ridiculed for asserting that the world was created a few thousand years ago. We, i.e. the scientists, say that we have a fossil record that "disproves" this. However, if you think about it, you cannot really *disprove this* with absolute certainty.

When Was the Universe Created?

Bertrand Russell pointed out that you cannot even *prove* that the universe wasn't created five minutes ago. The universe could have begun exactly 5 minutes before you read this line. All of your memories could have been initiated in that moment, including all the memories that give you the semblance of having had a much longer life. Similarly, all the fossil records that seem to document the passage of eons of time could have been planted exactly 5 minutes ago!

Why would a scientist be hesitant to embrace such a scenario? First of all, it makes no predictions. Prediction gives statements more power. We place more value and trust on statements that have predictive content, and this is not one of them. There are reasons to believe, moreover, that this scenario is very unlikely to have happened. All the information concerning the fossil histories and the memories of billions of people would have to be tuned exactly right so there would be no contradictions between them. Of course, this is possible *in principle*. But science puts more value on theories that do not require fine tuning, and the simplest explanation always wins! This is called *Occam's razor* principle. In science we embrace naturalness!

Science and Religion

A dominant view among some is to think of religion interfering with the practice of science. The most prominent example might be Galileo, who was persecuted by the church for his views and ordered to publicly refute them. But the notion of an inherent conflict between science and religion is not necessarily true; peaceful coexistence is possible. Isaac Newton, to take a prominent example, was an ardent religious scholar. Most of his writings were about Christianity and not science! In fact, his interest in studying nature was motivated by his religious views! He viewed himself as exposing God's laws rather than contradicting the notion of God. He hoped that in doing so, he would *help* people accept religion in light of the beautiful physical laws God has given us.

That is not to say that Newton accepted everything about religious institutions. For instance, he did not believe in the Trinity. And he certainly had his own interpretations of religion. Interestingly, Newton believed that God could, and did, interfere in physical reality. When some celestial objects were found not to behave as his equations predicted they should, Newton attributed this to God's intervention, which would allow exceptions to his physical laws! When compared to modern scientific views, Newton appears extreme in this regard. Yet it is also true that he was one of the greatest scientists of all time, and his religious convictions did not prevent him from compiling an amazing body of work that is still very much relevant today, more than 350 years after his accomplishments. Today we know that the heavenly object's orbit seemed to deviate from Newton's prediction not because there was an exception to his laws, but because he had not taken into account the attraction due to other nearby objects which was hard to see with telescopes.

Although some contemporary scientists believe that science and religion should be kept apart, to minimize conflict or interference, their domains may not be entirely separate. One area of possible overlap between science and religion involves our conception of what God looks like. This used to be a subject confined entirely to religious quarters, but given how abstract physics has become recently, a scientist might try to imagine how God could exist outside our own space-time, perhaps playing a role in higher dimensional multiverses. Drawing upon modern mathematical and theoretical physics, one might attempt to construct a consistent picture of God by going outside the universe. How this theory could be tested is another question altogether, but the exercise could represent a modern version of Newton's attempt to understand nature and the role that God plays in it.

The Origin of the Universe

Perhaps the most consistent overlap between science and religion concerns the origin of the universe. Almost every religion begins with a statement to the effect of "God created the universe." You might think that this comes into direct confrontation with science, but that is not necessarily the case. Scientists often make some assumptions about the initial conditions. In this case, God could be part of setting the initial conditions, and theory's predictions could take over just after the creation!

Recall that Einstein introduced the cosmological constant to keep his equations from describing an expanding universe. There was a Jesuit priest named Georges Lemaître who disagreed with this line of reasoning. As we already mentioned, Lemaître subscribed to the idea of a primordial atom and believed the universe should have started expanding from this primordial atom. He tried, moreover, to use Einstein's theory to demonstrate this effect. Einstein is said to have responded, "Your math is pretty good, but your physics is horrible." Maybe this criticism was accurate. Nonetheless, in Einstein's view, the universe had no beginning; it was eternal. In some versions of this story, Einstein seems to have accused Lemaître of trying to support Christian creationist mythology and appears to have taken his criticism too far.

We now believe there was a Big Bang, so in that sense the

priest was more correct than Einstein after all! What are we to make of this? Perhaps it was just pure luck, with Lemaître making one correct guess out of many thousands. On the other hand, it shows that a strong religious view is not necessarily incompatible with science. Scientists often ask questions that are shaped by religious views. Even if one is interested in pure science, *and perhaps for the sake of science*, we should not try to eliminate religion outright–and prevent it from entering the conversation–as it may serve as a source of inspiration. Another example of the beneficial role of religion concerns the rise of science during the height of the Islamic civilization about a thousand years ago, which has been attributed to the fact that many of the scientists at the time were inspired by the teachings of Islam and the Quran.

A closely related topic is the role of philosophy in science. Today, the pragmatism of American culture has influenced science greatly. Nowadays, scientists rarely talk about philosophy, and in fact many of them tend to look down on philosophy. This is probably rooted in American pragmatism. If you look back at the deliberations over quantum mechanics by Einstein, Heisenberg and other early practitioners, much of the discussion was philosophical in nature. This is rarely the case today among leading physicists. However, if you look deeply, you see that most scientists are influenced by philosophical principles, whether they admit it or not. Many scientists are perhaps unknowingly amateur philosophers! Philosophical principles can (and do) replace religious views for some scientists.

Einstein and Religion

Let us talk about Einstein and how he looked at religion. Simply put, Einstein was highly critical of conventional religion. In one letter, he wrote, "The word God is, for me, nothing more than the expression and product of human weaknesses, the Bible a collection of honorable, but still primitive legends, which are nevertheless pretty childish. No interpretation, no matter how subtle, can (for me) change this."

Here is another story: One evening, Einstein and his wife were at a dinner party when a guest expressed belief in astrology. Einstein ridiculed this as pure superstition. Another guest stepped in and went further, characterizing religion as superstition. The host intervened and said that even Einstein was religious. In response, Einstein said that he believed in a subtle, venerable structure in the laws of nature. This, he affirmed, was his religion.

Einstein said more in a 1936 letter written to a sixth grade student who asked him whether scientists prayed and, if so, what they prayed for. Einstein replied that "scientific research is based on the idea that everything that takes place is determined by the laws of nature." But Einstein also conceded that "our actual knowledge of these laws is only imperfect and fragmentary," adding that "belief in the existence of basic, all-embracing laws in nature also rests on a sort of faith [that] has been largely justified so far by the success of scientific research. But, on the other hand, every one who is seriously involved in the pursuit of science becomes convinced that a spirit is manifest in the laws of the Universe–a spirit vastly superior to that of man, and one in the face of which we with our modest powers must feel humble. In this way, the pursuit of science leads to a religious feeling of a special sort, which is indeed quite different from the religiosity of someone more naive."

Indeed Einstein viewed science with a religious bent. To see this more clearly one only has to recall Einstein's views about quantum mechanics and how he objected to probabilistic aspects of quantum mechanics by declaring to Bohr that "God does not play dice!"

Feynman and Religion

Feynman's views about religion are somewhat agnostic. For example, he believed that since we live in just this one little planet called Earth within a vast universe that has a huge number of galaxies and planets, why would God send his prophet only to our planet and neglect all the other planets? That hardly made sense to him, yet in no major religion do we hear about other planets and other beings and other prophets being sent there!

Feynman's views about physics are also somewhat iconoclastic. While many scientists believe that science will teach us something deep about the nature of the universe, Feynman was more focused on concrete problems. In contrast to the attitudes of most of his peers, Feynman viewed discoveries about the deep nature of the universe as a side effect and by no means his primary goal. This is in *contrast* to the prevailing attitudes of other researchers. In a way, many scientists may be, perhaps inadvertently, trying to *replace* religion with science as a paradigm for understanding the universe

It should be stressed, however, that Feynman did not dismiss religion outright. He admitted, instead, to having a longstanding interest in relations between science and religion. In a talk he gave in 1956, Feynman said that "many scientists *do* believe in both science and God in a perfectly consistent way. But this consistency, although possible, is not easy to maintain." One source of difficulty in trying to weld science and religion together, he said, owes to the fact "that it is imperative in science to doubt; it is absolutely necessary for progress in science to have uncertainty as a fundamental part of your inner nature... Nothing is certain or proved beyond all doubt. You investigate for curiosity, because it is *unknown*, not because you know the answer." And Feynman goes on to say that as you delve further in your investigation, "it is not that you are finding out the truth but that you are finding out that this or that is more or less likely."

This attitude, when applied to questions such as whether God exists, would, according to Feynman, prevent a scientist from attaining the "absolute certainty which some religious people have." Once the question of God's existence has been removed from the absolute, doubts pertaining to other aspects of religious doctrine may be raised. That was one of the reasons that Feynman felt it was difficult to be fully invested in both science and religion, and he chose science, while embracing the doubts that his choice entailed.

Hawking and Religion

Our modern view is that we have been able to push the frontiers of science to an enormous extent. Somehow, we have precise predictions about the beginning of the universe, which was presumably set into motion by the Big Bang that occurred approximately 13.8 billion years ago. Stephen Hawking described the Big Bang as a consequence of the laws of gravity, which did not need any help from a divine being. But what happened before that primordial blast? Does physics have anything to say about that?

Some scientists are pragmatists, saying that since this question cannot be the subject of experiments or applications, we should, therefore, ignore it. Other scientists have tried to think about this seriously, and Hawking was prominent among them. He raised the question: Can the universe arise from nothing, without any intervention? It turns out that there is a mathematical formalism, in the context of quantum gravity, which gives meaning to this statement. Hawking, together with James Hartle, drew up a quantum description (or wave function) of the universe, consisting of a path integral that sums up all possible past histories that could lead to a cosmos in the current state from nothing! The result of that analysis, and those of other physicists who have taken up this problem, suggest it is conceivable that this glorious universe–the only home we have ever known–could, in fact, have come from nothing!

Puzzle

Draw four straight lines passing through the 3×3 grid of points without lifting your pencil:

•   •   •

•   •   •

•   •   •

Solution

This is impossible if you stay within the box. The key is to go "outside the box." This is somewhat reminiscent of religious discussions in which it is said that we need to step above, or go outside our world, in order to see the answers–powers that are commonly attributed to God. (Although in this case all you really need is a sharpened, Number 2 pencil and perhaps a somewhat sharpened mind).

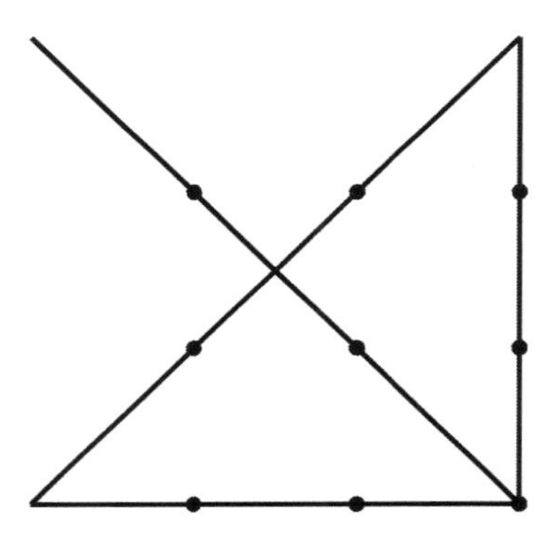

We can even do it with three lines, if the dots were big enough (the size of beachballs, perhaps, though this is meant to be a joke)!

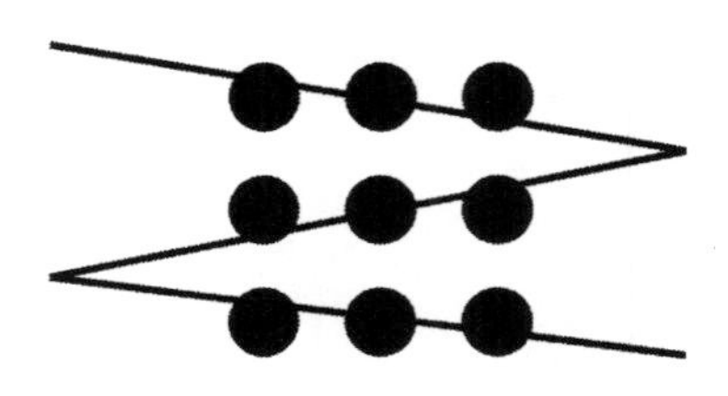

Puzzle

The owner of a house orders the gardener to plant 5 rows of 4 trees, but the gardener only has 10 trees. How can he do it?

Solution

Plant the trees in the shape of a star!

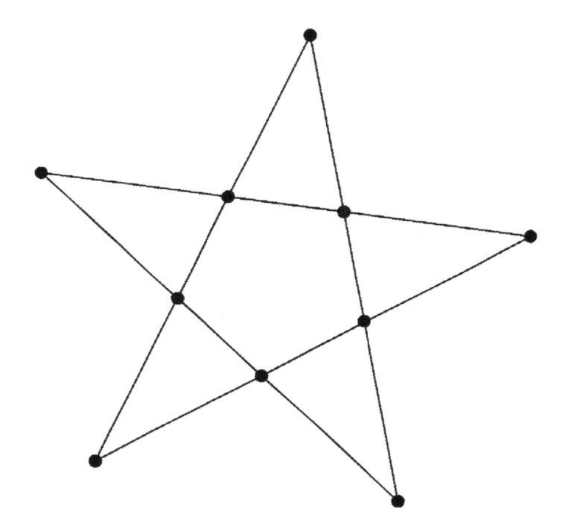

Pascal and Religion

Some people's attitude about religion is that one of the reasons to believe in religion is because it might just be true and, if it is true, you are better off believing than not. The French mathematician and philosopher, Blaise Pascal, justified this course of action with his famous wager, which was published in a book of his "Thoughts" (or Pensées) in 1670, eight years after Pascal died. The logic that Pascal applied was considered a milestone in probability theory and decision theory. His argument went roughly as follows: If God exists and you don't believe in him, then you are condemned to eternal hell. If God does not exist and you believe in him, then you get nothing bad out of that belief. Thus it is better to believe in God just in case he actually exists. So even if you believe the probability of God's existence is just one in ten thousand (or 0.0001), the expected return for believing is much better than that:

	believe	don't believe	probability
God exists	∞	$-\infty$	0.0001
God doesn't exist	$-\epsilon$	ϵ	0.9999
Expected value	∞	$-\infty$	

A similar logic can be found in a story about Niels Bohr, a pioneer in the fields of quantum theory and atomic structure. It is said that Bohr's uncle once put a horseshoe on the back of the house to keep evil spirits away. Some asked him why he was motivated to do that. "Surely you don't believe in such superstitious stuff?" he was asked. To which Bohr's uncle replied: "No, of course not. But they say it works even if you don't believe in it!"

Causality and God

The idea of *causality* may explain why some people believe in God. An old theological argument holds that since everything has a cause and a reason it leads to a chain whose origin–or so the reasoning goes– must be God. There are some standard rebuttals: Who created God? The usual reply would be that God is the only entity that needs no creator to exist. But then the criticism would be that if something exists that needs no creator, why can't it be the whole universe?

This discussion indicates that our actions and beliefs are not dictated solely by causality or other intellectual constructs. We are human beings, and at the deepest level, we are not strictly logical. We are emotional, intuitive, sometimes inspired, and sometimes just wrong. Scientists, of course, cannot divorce themselves from being human, nor should they try. We believe in science not just for pragmatic reasons. Some of us are motivated by the pursuit of some sort of absolute truth–or as absolute as we can get, given the limitations we bring to the problem and the intrinsic uncertainties of the world we inhabit. The existence of such a high-minded and, to some extent, unrealistic goal is not rationally defensible, nor is it defensibly rational. In fact, it is somewhat akin to a religious belief.

Puzzle

Consider a rectangle of length a and width b, tiled by smaller rectangles having the property that each small rectangle has integral width or integral height, meaning that at least one of them–the width or the height–is an integer (see Fig.79). Whether it is the height or width that is integral may differ between the tiles. Prove that the larger rectangle enjoys the same property.

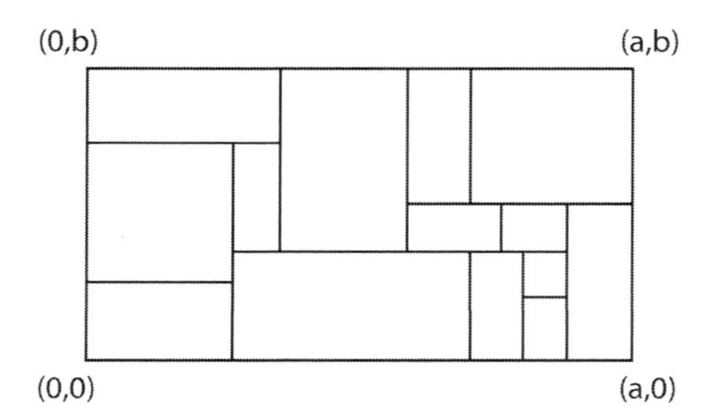

FIGURE 79. A big rectangle made of smaller rectangles all having integral height or width will inherit this property.

This puzzle demonstrates that amazing global properties can be derived from seemingly random local structure. So, in this context, one does not need a "God" or other grand overseer to arrange for the global structure by hand, because it emerges, naturally, from local properties.

Solution

Anchor one vertex of the large rectangle at the origin, and the others are $(a, 0), (0, b)$, and (a, b). One elegant solution would be to consider integrating $\sin(2\pi x + \phi)\sin(2\pi y + \theta)$ over the rectangle for arbitrary angles θ and ϕ:

$$\int_0^b \int_0^a \sin(2\pi x + \phi)\sin(2\pi y + \theta)dxdy =$$

$$= \frac{1}{4\pi^2}\left(\cos(2\pi a + \phi)) - \cos(\phi)\right)\left(\cos(2\pi b + \theta) - \cos(\theta)\right)$$

For each smaller rectangle, the integral is zero because it is a product of two factors at least one of which is zero due to integral length. Therefore, the integral is zero over the whole rectangle, and this is only possible if the length of at least one side is an integer. (Given that the cosine is a periodic function that is unchanged by integer multiples of 2π, if a and b are integers, then both terms on the right-hand side of the equation equal zero, and the integral would also be zero).

This solution can also be solved by a parity argument, which means using a mod 2 argument. Let N be the number of pairs (R, p) such that p is an integral point (i.e., with x and y being integers) and is also a vertex of the rectangle R. For each rectangle R in the subdivision, its number of integral vertices (i.e., vertices that are themselves integral points) is even (because at least one side is integral) so each rectangle R contributes an even number to N. Therefore, N must be even.

Another way to count N is to ask for each point p, how many rectangles include it as a vertex and then add up the total to get N. For each point p, the number of rectangles incident at it is even (see Fig.80), *except* for the four corners of the large rectangle (see the graphs below). Since the total number N is even, an even number of the corners of the large rectangle are integral. Since we put one corner of the large rectangle on the origin (which has integral coordinates), we must have at least one more integral corner.

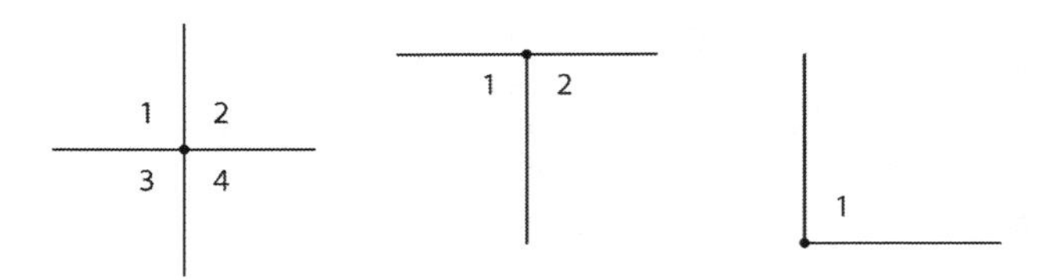

FIGURE 80. Counting the number of rectangles that could be incident to a point p. The number are all even except at the vertices of the large rectangle.

This puzzle can be also extended to 3D (and higher dimensions), with length, width, or height integers.

11. **Duality**

The topic of duality, which is becoming increasingly important in both physics and mathematics, offers a nice way to tie together many of the topics we have covered so far. Nowadays, it is a common occurrence in mathematics and physics that one might be trying to answer a complicated question that is somehow equivalent, or "dual," to a much easier question. And the answer to the easier question can almost trivialize the complicated one, showing that it is not nearly so difficult as was initially thought. Good puzzles are like that too. All it takes to solve them, sometimes, is a shift in perspective. The key lies in knowing how, and in which way, to shift. So, in this sense, the notion of seeking dualities seems to be ingrained in us, as we are naturally driven to seek the easiest way of solving a problem.

Over the years, we have learned, for instance, that five different versions of string theory are mathematically equivalent to each other; that 10-dimensional string theories are dual to 11-dimensional M-theory–a gravitational theory of membranes; and that string theories are also dual to some lower-dimensional quantum field theories. Which version one chooses depends on the problem at hand.

The AdS/CFT correspondence,[36] to pick another example, is

[36]This was originally proposed by Juan Maldacena and has been developed further by many physicists.

a duality between a gravitational theory governing a region of space-time and a gravity-free quantum field theory that pertains to just the boundary of the same region. This correspondence was discovered more than 20 years ago and it continues to generate new insights, along with many surprises. "People keep finding new facets of dualities," the physicist Edward Witten has said. "Dualities are interesting because they frequently answer questions that are otherwise out of reach. For example, you might have spent years pondering a quantum theory, and you understand what happens when the quantum effects are small. But textbooks don't tell you what to do if the quantum effects are big; you're generally in trouble if you want to know that. Dualities frequently answer such questions. They give you another description, and the questions you can answer in one description are different than the questions you can answer in a different description."

Dualities represent symmetries, in a sense, as well as *highly* "nontrivial equivalencies," and by the latter I mean things that are not obviously true. Dualities have recently allowed us to solve extremely complicated problems in mathematics and physics. Curiously, we do not know why many of these dualities are true. They are a puzzle solver's dream, but they are also puzzles in and of themselves. When we solve a puzzle, we normally have some understanding of how we arrived at the answer. However, when we rely on a duality to solve a puzzle, we often do not understand how it works. That can make for a pretty awesome trick, but the fact that the explanation is beyond us can also be rather embarrassing and frustrating!

Two Mathematical Examples

Suppose we are in some nice D-dimensional space, and we are considering objects A_ℓ of dimension $\ell = 0, \ldots, D$. Then there is a duality of A_ℓ with $B_{\tilde{\ell}}$, where $\tilde{\ell} = D - \ell$. In other words, the duality takes objects of dimension ℓ to objects of dimension $\tilde{\ell}$. For example, for $D = 2$, a 0-dimensional object (a point) transforms into a 2-dimensional object (such as a triangle) and vice versa, and lines are dual to lines as seen below. This is a manifestation of "Poincaré" duality.

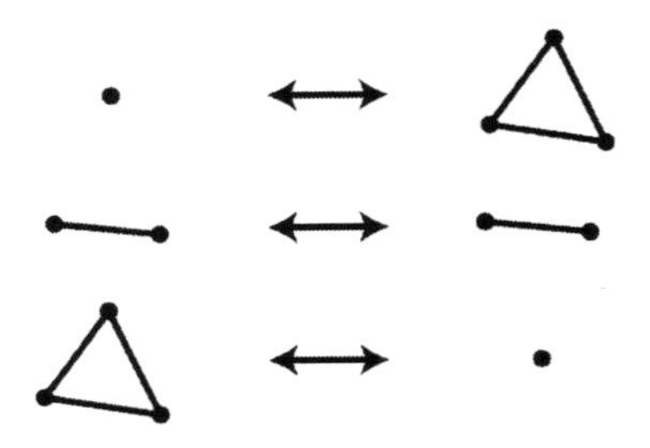

In the 2-dimensional example in Fig.81, we have points exchanged with faces, and lines dual to lines. Every statement about a triangulation has a dual analog, as can be seen below.

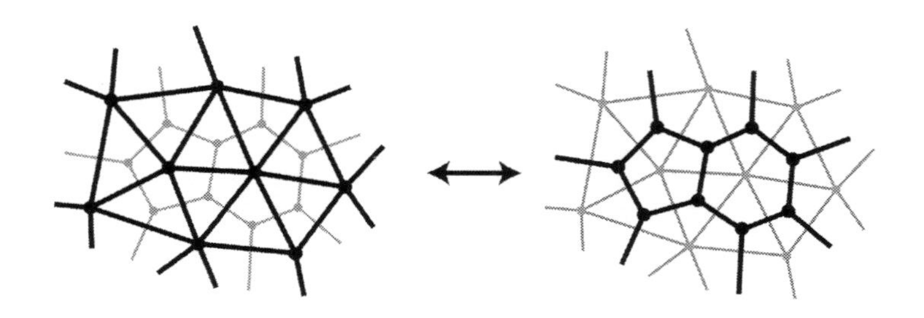

FIGURE 81. An example of Poincaré duality in 2 dimensions

As another example of mathematical duality, suppose we are trying to solve the differential equation

$$\sum_{n=0}^{N} a_n \frac{d^n f}{dx^n}(x) = 0.$$

This looks horribly complex, but suppose we try a function of the form $f(x) = e^{px}$. Then the $\frac{d}{dx}$ operation becomes multiplication by p, and the differential equation becomes

$$\sum_{n=0}^{N} a_n p^n = 0.$$

So we have taken a scary looking differential equation and turned it into a polynomial equation. That can be quite useful because, generally speaking, polynomial equations are much easier to solve than differential equations. The transformation we made is a simple example of the Fourier transform. In a Fourier transform, we write a function $f(x)$ as a sum (or integral) of complex exponentials: $f(x) = \sum c_\alpha e^{i\alpha x}$. The Fourier transform is not an approximation of the function; it is a dual way of describing it. And by Fourier transforming $f(x)$ from its usual space to what is called its frequency space, we have managed to trivialize a difficult problem.

These are dualities that we can formulate mathematically and prove rigorously. But recently in physics, we have discovered dualities that are more mysterious and far more powerful than the Fourier transform. From a mathematical viewpoint, these dualities have not yet found a clear explanation. Imagine if you were trying to solve an exotic differential equation, and you had a black box method that would produce an answer, which you could then check to see if it works. There are many such statements in modern physics today. It is like having a magical code for looking at problems, something akin to a master puzzle solver. We do not know how or why they work, but they do. To put this in perspective, these methods, based on dualities, can solve previously unsolved mathematical problems! In other words, they can give us the right answer without giving a deep explanation for how we got there.

Duality in Quantum Mechanics

Let us remark that the Fourier transform was critical to the development of the particle-wave duality in quantum mechanics.

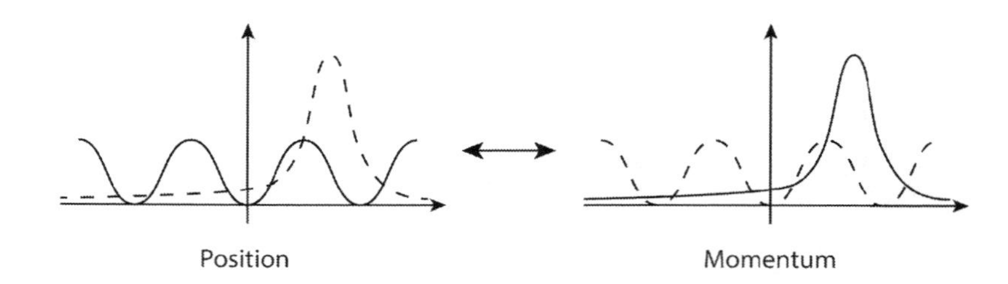

The Fourier transform of position space is momentum space. It takes distributions that are highly concentrated in one space (low uncertainty in position) into distributions that are highly dispersed in another. Stationary objects are more like particles, resembling probability functions that are peaked near a specific point in space. Moving objects are more like waves, spread out in position space, which can be viewed as probability distributions that are peaked near a point in momentum space (the Fourier transform of position space). The more peaked a distribution is in one space, the more spread out it is in its dual Fourier transform. The precise relation between position and momentum is given by Heisenberg's uncertainty principle:

$$\Delta x \Delta p \geq \hbar/2.$$

Why do these dualities emerge? Why are there more pictures than one? We do not have a good explanation for this phenomena, but the existence of dualities appears to be a deep, ineluctable fact of nature.

Maxwell's Theory

In the theory of electromagnetism, we consider electric and magnetic fields, denoted by $\vec{E}$ and $\vec{B}$. An electric charge q_e induces an electric field. One might think that a magnetic field is induced, similarly, by a magnetic charge q_m, but we have never discovered such things in nature–particles, or magnetic monopoles, containing an isolated unit of magnetic charge, a north pole, for instance, without a south pole. Dirac showed, however, that if magnetic monopoles do exist, then electric charge is necessarily quantized. Based on arguments from quantum formulation of gravity we expect that there are magnetic monopoles. Moreover, they would arise naturally in the context of the unification of electromagnetic forces with the other forces we discussed before.

Maxwell's equations enjoy an interesting symmetry between the electric field and magnetic field:

$$\begin{cases} \vec{E} \mapsto \vec{B} \\ \vec{B} \mapsto -\vec{E} \\ q_e \mapsto q_m = 1/q_e \end{cases} .$$

Of course, if there are no q_m's, or if their masses are different for electrically charged states, then this is not a symmetry. However, even if magnetic monopoles do not exist, it does make sense in empty space where there are no charged particles. In this setting of empty space or a vacuum, Maxwell's theory has an amazing symmetry: If you take Maxwell's equations and replace the electric field with the magnetic field, and replace the magnetic field with the negative of the electric field (which is what the above arrows signify), the equations are unchanged. But remember, this remarkable symmetry only applies to empty space.

In quantum theory, the situation becomes more complicated. Quantum effects destroy the symmetry between the electric and magnetic fields. In order to quantize Maxwell's theory, you need to know the quantum charge

$$\frac{e^2}{\hbar c} \approx \frac{1}{137}.$$

This number controls the quantum fluctuations: the bigger it is, the bigger the quantum corrections are. Since this number is pretty small–below the one percent level–quantum effects are not a dominant factor in our daily lives.

In quantum theory, the symmetry involving the electric and magnetic fields more or less *inverts* this quantity

$$\frac{e^2}{\hbar c} \mapsto \frac{\hbar c}{e^2} \approx 137.$$

Roughly speaking, the quantity representing the strength of electric interactions inverts to become the strength of magnetic interactions, which is much larger. That's because $q_e q_m \approx 1$. The electric interaction is therefore called weakly coupled and the magnetic interaction is strongly coupled. The stronger the magnetic interaction, moreover, the weaker the electric interaction and vice versa.

However, as noted previously, the symmetry between electric and magnetic interactions does not work due to the absence of magnetic monopoles (as well as quantum effects) in Maxwell's theory. But there is a modification of Maxwell's theory for which this symmetry does work, which I will now explain.

You can take the theory $(\vec{E}, \vec{B}, \ldots)$ and, instead, generalize them to matrices. This is called the "non-abelian" version of Maxwell's theory, because, unlike numbers, matrices do not commute and thus form a non-abelian group.

Let us delve into the math a bit more. In Maxwell's theory, the components of fields are numbers that can be viewed as 1×1 matrices, based on what is called a $U(1)$ gauge symmetry. A 1×1 matrix is, again, equivalent to a number, but expressing the number in that form allows one to scale up the picture more readily. If the components of electric and magnetic fields were instead $N \times N$ (Hermitian) matrices, we would get what is called the $U(N)$ gauge symmetry.

The strong/weak duality that we have mentioned still does not quite work, unless we add enough fermions to make the theory supersymmetric, as we have discussed before. That tames the quantum fluctuations and restores the duality between the electric and magnetic fields from the classical theory.

This leads to a duality that we do not fully understand–the validity of which we have evidence for–though we have no idea, at present, how to prove it.[37] This turns out to be connected to questions that mathematicians are independently interested in, as part of the so-called geometric Langlands program, which is related to problems in number theory. The strong/weak duality connects math and physics in this and other ways. Physicists have checked this duality in many, many nontrivial cases, and it works, but we still cannot explain from first principles why it works. While the statements of physicists can be expressed in concrete mathematical pieces, we will not be able to unify them until we have a complete, mathematically rigorous understanding of a quantum field theory–something that is still lacking, nearly a century into the quantum era.

The strong/weak duality between magnetic and electric interactions is an example of an S-duality whose most studied version is the S-duality of $N = 4$ supersymmetric $U(N)$ Yang-Mills theory.

[37]The only case of this that we can actually prove is the $U(1)$ case which can be proven using Fourier transform in the infinite dimensional space of quantum fields.

Duality in String Theory

String theory has laid bare many powerful and amazing dualities, thereby illustrating the importance of dualities in physics. The duality between weak and strong couplings of electric and magnetic charges, as we shall see, can be translated into the geometric language of string theory. In that setting, we do not consider only four-dimensional space-time ($\mathbb{R}^4$) but must also include other higher dimensional geometries needed to ensure the theory's consistency. The relevant part of the geometry for the problem at hand turns out to be 6-dimensional. The relevant 6-dimensional geometry is taken to be a product of the usual Minkowski space-time $\mathbb{R}^4$ and a 2-dimensional torus (i.e., $4 + 2 = 6$). This additional torus could, for example, have side lengths ℓ_1 and ℓ_2, whose ratio is

$$\frac{\ell_2}{\ell_1} = e^2.$$

The side lengths we just referred to, ℓ_1 and ℓ_2, can be thought of as the lengths of a rectangular piece of paper. The paper, in turn, can be rolled up into a cylinder and the ends of the cylinder can be attached to each other to form a donut or torus as in Fig.82.

In the context of string theory, the $E \leftrightarrow B$ symmetry for $U(N)$ theory come about by wrapping N 6-dimensional objects on a 2-dimensional torus, where the resulting 4-dimensional theory enjoys this symmetry. In that formulation, this symmetry comes from viewing the torus as a rectangle (Fig.82). If we rotate the coordinates (counter-clockwise) by $\pi/2$, then we will have the transformation

$$\begin{cases} x \mapsto y \\ y \mapsto -x. \end{cases}$$

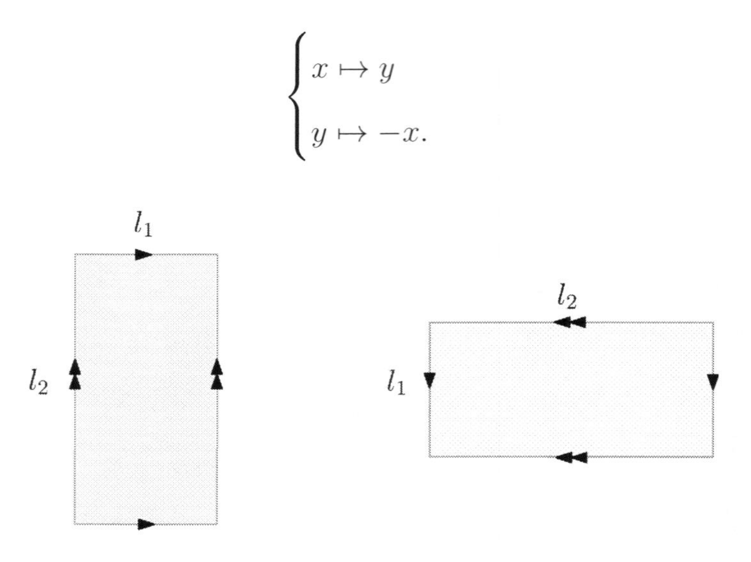

FIGURE 82. Electric/magnetic duality can be viewed as a 90 degree rotation on a 2-dimensional torus.

This operation changes what we mean by e^2 because the two sides have switched roles and instead we get $\frac{\ell_1}{\ell_2} = 1/e^2$ which we identify as the magnetic charge squared. The duality between electricity and magnetism, when translated into string theory, becomes this trivial observation about the fact that there are no distinguished sides of a torus. By that we mean that for the original rectangle that gave rise to this torus, it does not matter which side you call horizontal or vertical. The choice is completely arbitrary: $\frac{\ell_1}{\ell_2}$ and $\frac{\ell_2}{\ell_1}$ give rise to the same physics.

T-duality

The simplest, though one of the most far-reaching, dualities in string theory is "T-duality." To get a sense of what this is, take a periodic interval of length L, which you could alternatively think of as a circle. You could also consider a periodic version of square of side L leading to a torus. Or a 3-dimensional version of this periodic box also known as a 3-dimensional torus. Imagine that this box is the universe and shrink L. It gets smaller and smaller, and seemingly more compressed as in Fig.83.

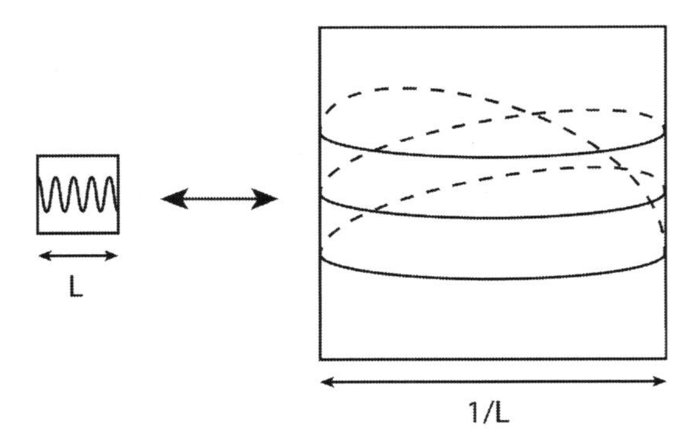

FIGURE 83. A small universe of length L is dual to a large universe of length $1/L$ where the moving strings are dual to winding strings.

But in string theory something remarkable happens: When you continue shrinking it below the string length, it eventually behaves *as if* it is "expanding." There is a length inversion duality: the universe of length L in string units is *dual* to a universe of length $\frac{1}{L}$.

This sounds crazy, but it is one of the dualities that we can actually prove! The energy of a particle in a box is quantized, and given by $E \sim \frac{n}{L}$ if the length is L. This is the energy typically associated with harmonics in a box, and n/L can be viewed as the momentum of the string's center of mass. Note that this is not symmetric under $L \to 1/L$. As the length gets small, the energy of these modes becomes huge. String theory brings an additional ingredient to the situation–the winding of strings around the box. This is characterized by its winding number, so its energy is proportional to L: $E \sim mL$ (because you have to do work to stretch the string), where m is the winding number. Now, observe that if you take $L \mapsto 1/L$ and exchange m and n (i.e., exchange the winding string modes with momentum of center of mass of strings) you get the same spectrum for the theory. In other words, as far as energy is concerned we cannot distinguish a box of size L and a box of size $1/L$.

In fact, more broadly, what we are saying is that you cannot distinguish a universe of size L from one of size $1/L$ in string theory. Not knowing about string theory, Einstein may well have disagreed with this conclusion. He might have said that if you want to measure the length, then just shine a light and measure how long it takes to travel the length of the universe. This would give an unambiguous and definite length. So how might we resolve such a difference of opinion? Well, in string theory, there are two kinds of light–our usual conception of light (comprised of momentum modes) and one comprised of winding strings. If you measure distance with the usual kind of light, involving ordinary photons, you get L as the length. But if you use the other, dual version of light, involving winding photons, then you will see the other length, $1/L$! This tells us that distance is not a fundamental notion of string theory. You might also wonder whether we could see this other light in, say, our flashlight at home. The answer to that is no because its energy is proportional to the length of the universe, and it therefore would take an astronomical amount of energy to create it (not practical given current battery technology)!

Calabi-Yau Manifolds and Mirror Symmetries [38]

T-duality can give rise to new dualities. Specifically, Calabi-Yau manifolds are very special kinds of manifolds, each possessing a dual or "mirror" manifold that has a different complex topology. Although these two Calabi-Yau manifolds would be

[38]For a very readable account of this see Brian Greene's *The Elegant Universe.*

classified as distinct from a mathematics standpoint, that's not necessarily the case in string theory where manifolds of different topology can nevertheless give rise to the same physics. This generalization of T-duality is called *mirror symmetry.*

Dualities turn questions into dual questions, meaning that for every question you can ask in one framework, there is a dual question that can be asked in a dual framework. What is an example of an interesting question in one setting that gets reformulated as a dual question in another ? Well, in order to compute physical interactions in string theory, we start with 10 dimensions and reduce the theory to 4 dimensions by assuming that the remaining dimensions are curled up and hidden on a tiny, 6-dimensional space, a typical example of it is known as a "Calabi-Yau manifold." We then need to to calculate the number of minimal area spheres that can be placed inside that manifold, which can be a very difficult task that is, in some cases, beyond us. Because of the duality, however, we can answer the same question by computing some simple integrals on the mirror Calabi-Yau–in this way replacing a vexing problem with a much easier one. Using this method, physicists computed the number of minimal area spheres for spheres of different degrees–each degree corresponding to the winding number of the sphere, or the number of ways a sphere can wrap around the space. Mathematicians had previously worked out this problem for degrees 1 and 2, and their answer agreed with the number that physicists had obtained. However, by using mirror symmetry, physicists were able to determine the answer not just for degrees 1 and 2 but for *any* degree.

Mathematicians had originally tried to get these numbers using traditional methods and, after some hard work, they obtained a solution to the degree 3 problem. Their number, however, disagreed with that arrived at by physicists through an approach involving mirror symmetry. Many people assumed the string theorists had gotten it wrong, but the mathematicians later discovered an error in their own work. After redoing their calculations, they confirmed the physicists' computations. This gave us further confidence in using dualities to solve difficult problems in physics and math, because this approach yielded reliable predictions that could not be obtained by other known means.

We can also ask the question, not only regarding the number of minimal spheres that can be placed on a Calabi-Yau, but also about how many minimal g holed surfaces are there in the Calabi-Yau manifolds (where $g = 0$ is the case corresponding to the sphere). g is called the genus of the surface. More than twenty-five years ago, physicists computed these numbers for genus 1 and 2, extending this all the way up to $g = 49$ about a decade later. Without exploiting mirror symmetry, mathematicians have so far only reproduced the $g = 1$ case.[39] This demonstrates the power of these mysterious dualities. Even though it is not yet deeply understood mathematically, mirror symmetry is exciting to mathematicians because they can look to physics to address mathematical problems, as well as to motivate new mathematics.

Mirror symmetry, which came to the attention of researchers through the exploration of string theory, is among the easier dualities for physicists to take advantage of. The S-duality for the non-abelian version of Maxwell's equations which we discussed earlier, on the other hand, is far more complicated to understand.

Let us try to give analogies through simple puzzles that might help convey something about the nature of the string theory dualities and how to think about them.

Puzzle

Consider a square board of size 1000×1000 cm made of 10^6 square grids of size 1×1 cm. You are given square shaped bricks of size 1×1 cm which you are instructed to place on the square grids in a non-overlapping fashion). Suppose you are

[39]These results from physics are viewed by mathematicians as conjectures, owing to different standards of rigor. Often what physicists call "established results" mathematicians would call "conjectures from physicists."

given 999,990 such bricks. Find all the possible ways you can place them on the board.

Solution

The answer is $\binom{1000000}{999990} = \binom{10^6}{10}$. We are using the symmetry of the binomial coefficient

$$\binom{n}{k} = \binom{n}{n-k}.$$

That is a symmetry, because we get the same answer, regardless of which side of the equation we use. But it is also a kind of duality. We can view the tiled squares as untiled and the untiled squares as tiled; the two situations are completely symmetric. In other words, each way to place squares on the board can also be thought of as a way *not* to place squares on the remaining squares. So, the answer is the same as the number of ways to put 10 bricks on the board, and that is a much simpler, more manageable problem to work out.

In some sense, this embodies the unproven though intriguing idea that there cannot be infinitely complicated physical theories. By including more and more squares, the problem initially becomes more complicated, but then it eventually starts getting easier and easier. Maximum complexity is not achieved when the parameters attain their biggest possible value, because if we were given 10^6 bricks to start with, there would be only one way to place them on the board. In this case, maximum complexity occurs when $k = n - k$, i.e., when $k = n/2$, which happens when we have $\frac{1}{2} \times 10^6$ bricks. This is also reflected in the case of T-duality where the smallest and thus physically most complicated effective length we get is when $L = \frac{1}{L}$, i.e., when $L = 1$.

Puzzle

You are given a meter long stick and 20 ants that you are to place on the meter stick at $t = 0$. You can place the ants anywhere on the stick and instruct them initially to go left or right on the stick, moving at the speed of 1 meter per minute. Each time two ants collide they simply reverse directions and continue moving with the same speed. When the ants get to the end of the stick, they simply fall off. Your task: Where do you put the ants, and which directions should they initially move so as to maximize the time before the last ant falls off the stick?

Solution

Instead of viewing the ants as reflecting off of each other, view them as walking through each other by dualizing the ants and changing their identities as they collide (See Fig.84). This does not change the positions of where the ants are, if you do not keep track of their identities. Then it is clear that one can ignore collisions entirely, in which case any solution where at least one ant begins at the very end works and gives the longest time of one minute, before the last ant falls off the stick.

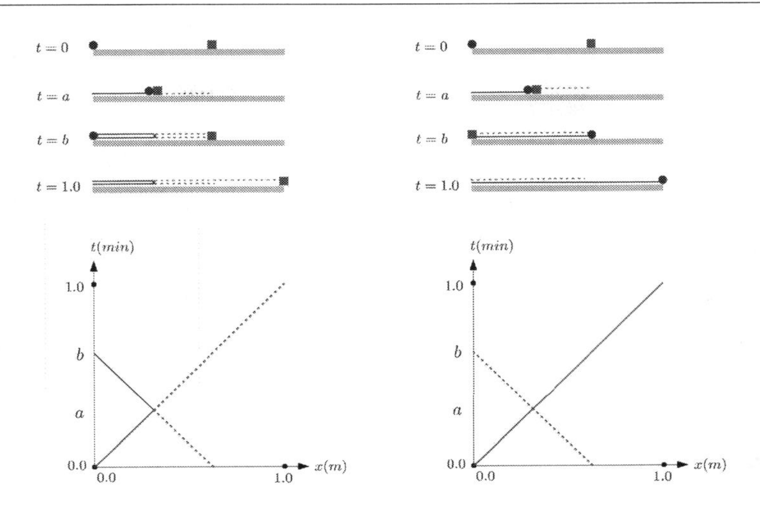

FIGURE 84. Duality of colliding ants: exchanging identities after collision.

Puzzle

Suppose that points A and B are separated by 100 km. There is a person riding a bicycle from A to B at 1 km/hour and a car driving from B towards A at 100 km/hour. Whenever the car meets the bicycle, it turns around; when it gets to the end, it turns around again, and continues this until the bicyclist gets to B. What's the distance that the car travels?

Solution

Again, it's possible to give a more complicated argument than necessary by summing the distances for each leg of the journey. However, the simplest procedure is to calculate the time it takes for the bicycle to get to B, and multiply the number of hours by the car's speed (100 km/hour) to determine the distance it travels during this time. In other words, it is easier to calculate the dual time, measured by the bicycle rider, which is the same as that measured by the traveling car.

Other Dualities: Geometry and Force

According to string theory, the universe is composed of the product of 4-dimensional space-time, $\mathbb{R}^4$, and a compact space which is 6-dimensional. Physics can be "geometrically engineered," in the sense that the forces and particles of nature can be interpreted in terms of the geometry of this compact 6 dimensional manifold. Just as Einstein told us that gravity is a manifestation of the curvature, or geometry, of space-time, string theorists contend that much of the physics we see is dictated by the shape, or geometry, of the hidden 6-dimensional manifold found at every point in space. The theory holds that we live in $(3 + 1)$ space-time, and maybe 6 (or 7) hidden dimensions. Where are the hidden dimensions? Well, a telephone wire, as seen from far away, looks like 1D as in Fig.85, but we can see it is 2D upon closer inspection. Macroscopic space is 3D, but just as in the case of the telephone wire, there could be additional hidden dimensions that are too small for us to see.

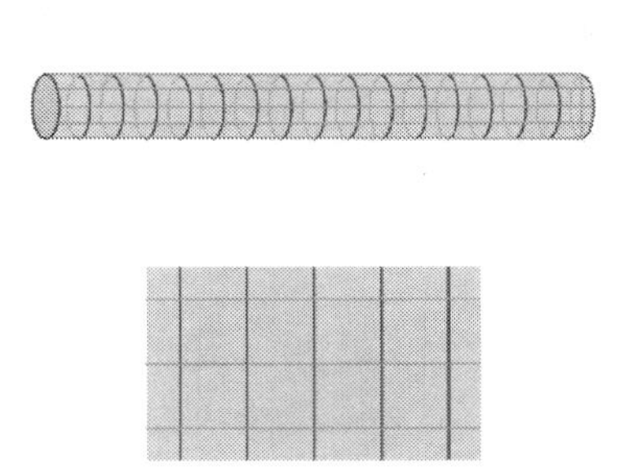

FIGURE 85. Extra dimensions can appear upon closer inspection.

In string theory, there are 4 macroscopic space-time dimensions and 6 compactified ones as in Fig.86.

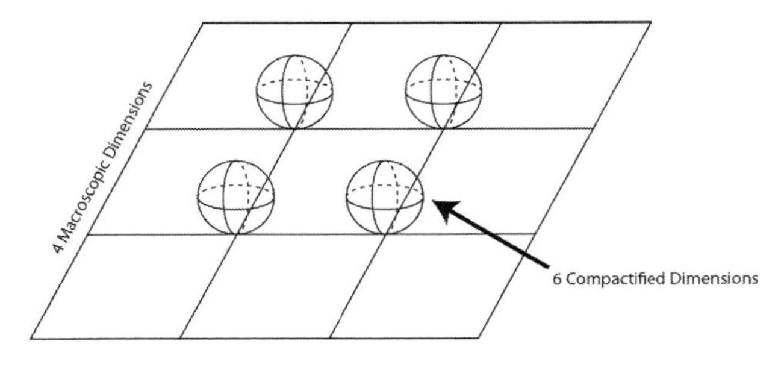

FIGURE 86. At each point in macroscopic space there are tiny curled up spaces.

What do these tiny, extra spaces look like? How do they impact what we observe in space-time? This series of questions has led to the aforementioned notion of the geometric engineering of physics. Based on how the shape and size of the internal 6-dimensional space looks, the observed physics will be different, with different particle masses and force strengths and so forth. For example, suppose we want to describe the geometry of the strong interaction that governs the forces between quarks. What we would need to do for the tiny internal space is to have it encompass two spheres touching at a point as in Fig.87 (leading to a non-abelian gauge force for $SU(3)$). The way that physics and geometry relate to each other–in this and other instances–is really quite magical, although too complicated to explain here.

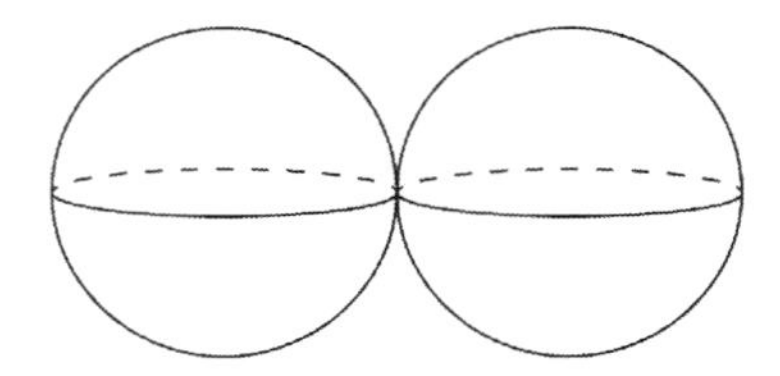

FIGURE 87. Two tiny spheres touching at a point leads to the physics of strong forces between quarks.

Strictly speaking we have to consider the limit where the area of the spheres approaches zero.

Puzzle

Consider 4 ants on a plane moving at constant velocities in different directions. We will label them 1,2,3, and 4. Suppose we are told that all of the ants will collide in pairs as they move, except for ants 1 and 2; we are not told whether they will collide with each other or not. But given this set-up, can we say for certain that 1 and 2 will collide?

Solution

They necessarily will. To see that, and see the power of going to higher dimensions, consider adding time to this scenario. In other words, consider the space-time given by (x, y, t) where (x, y) denotes the points on the plane the ants are on, and t denotes the time. If we consider the trajectory of each ant in this space-time we get a line, which is called the world-line. Since each ant is going at a constant velocity, we conclude that each ant must have a straight line as its world-line in the (x, y, t) space. The fact that two ants collide implies that their world-lines must intersect because at some time t they must both be at the same position (x, y). In particular, the word-lines of ants 1 and 3, as well as ants 1 and 4 intersect. Also the world-lines of ants 2 and 3, as well as ants 2 and 4 intersect. This implies that the three world-lines of ants 1,3,4 form a plane and so do the world-lines of ants 2,3,4 (see Fig.88). But the world-lines of ants 3 and 4, given that they intersect, define a single plane. That means that the world-lines of ants 1 and 2 are on that same plane. Given that ants 1 and 2 go in different directions (and are, therefore, not moving parallel to each other), and that their world-lines are on the same plane, their world-lines must cross, meaning that the ants will collide somewhere on the plane.

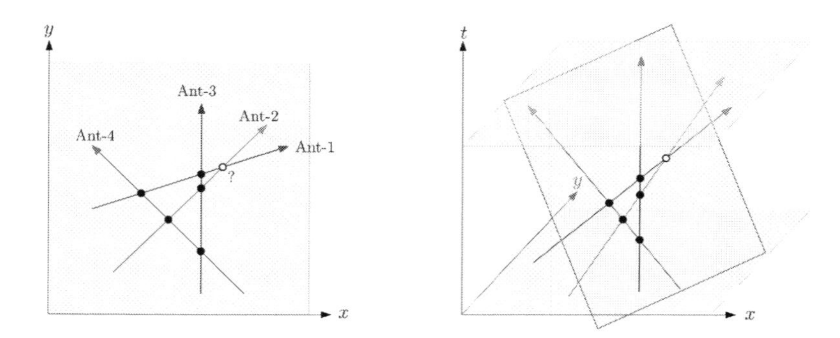

FIGURE 88. The abstraction leading to viewing time as an extra coordinate leads to the solution of the puzzle by showing world-lines of ants 1,3,4 and 2,3,4 are on the same plane in space-time. This leads to the conclusion that the 1,2 world-lines are on the same plane.

Duality in Black Holes

Hawking revealed that black holes have an extraordinarily high entropy, and he (using a work by Bekenstein) showed that the entropy is proportional to the area of the event horizon. But where does this entropy come from? What are its microscopic ingredients? My colleague Andy Strominger and I were able to get an exact solution for the entropy, or internal degrees of freedom, of a black hole using a dual description in string theory. The string theory calculation involved counting the number of spheres–around which membranes, or "D-branes," are wrapped–that can fit inside a 6-dimensional Calabi-Yau manifold. This approach yielded the same answer as the Bekenstein-Hawking formula, while offering a detailed internal picture that showed how black holes could have such high entropies. This was a notable achievement for string theory as well as a testament to the power of dualities: counting mathematical objects inside a Calabi-Yau miraculously gave the same number as that deduced from the horizon area of a black hole.

Many significant dualities that come up in string theory involve a change in geometry referred to as a "geometric transition" (see Fig.89). Imagine, for instance, a horizontal torus like a donut sitting on a table. Now picture a circle aligned vertically with its base on the table. Suppose we shrink the circle down to a point, pinching the donut in the process until it comes undone and opens up, thereby becoming topologically equivalent to a sphere. This provides a simple picture of a geometric transition, though it does not convey their importance to string theory. One key example of this is the notion of holography, which we will discuss next.

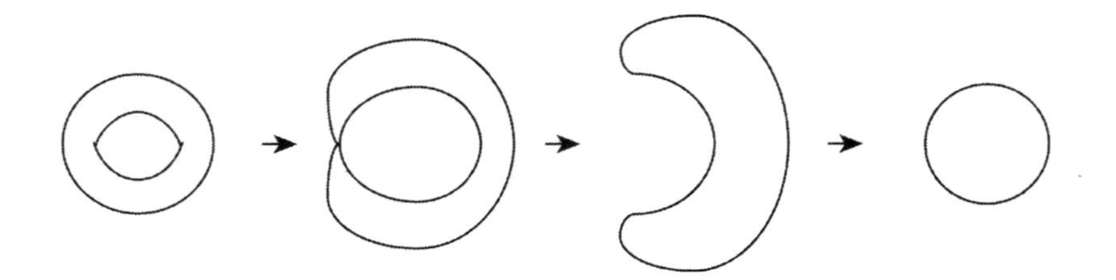

FIGURE 89. Example of a geometric transition: a torus transits into a sphere.

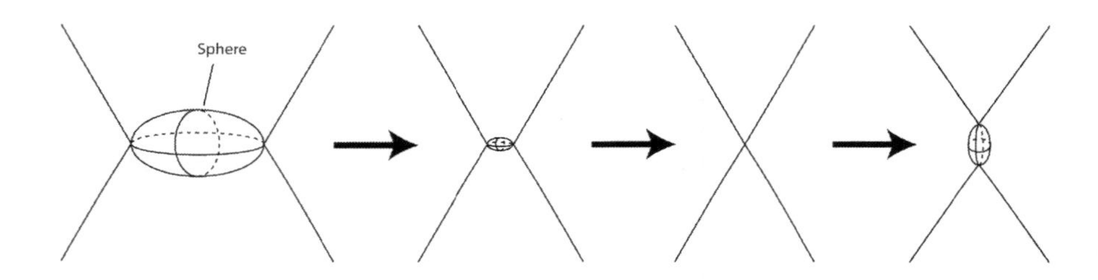

FIGURE 90. A sphere shrinks and another sphere opens up: a geometric model for quark confinement.

Quark confinement, the fact that quarks are strongly bound inside the nuclei, can be represented by a geometric transition, where a sphere shrinks and another sphere grows instead, as in Fig.90.

Holography

You are probably familiar with holographic cards, which have 2D images that give the impression of 3D. We also mentioned holography in the context of black holes, which can be completely characterized if we know the surface area of the event horizon rather than their volume.

Suppose you take a photon, classically represented by a 1×1 matrix, and represent it instead by an $N \times N$ matrix. This is the subject of Yang-Mills theory. If you take $N \gg 1$ really large, and interpret it in $\mathbb{R}^4$, then you get the theory of gravity in 5 dimensions! This is holography, as well as an example of the AdS/CFT duality. AdS/CFT can be described in simpler language: All the information about a 5-dimensional space-time–as described by string theory, a theory that includes gravity–is completely encoded on the 4-dimensional boundary of that space-time, as described by a quantum field theory that does not include gravity. Remarkably, these two pictures–involving different dimensional space-times, one including gravity and one not–turn out to be equivalent. This is not only a startling fact but also a very useful one, as it has fueled interesting work in theoretical physics over the past two decades.

Wigner Semicircle Law

Consider the Gaussian distribution, given by the density function

$$f(x) \propto exp(-x^2/g).$$

Wigner asked: What if you consider a higher-dimensional (matrix) analogue? That is, x is now replaced by X, a symmetric

$N \times N$ matrix, with each entry being a random variable. Now let us consider its eigenvalues. Generically, there will be N of them. Wigner found that if $N \gg 1$ and $g << 1$ with Ng fixed, then the eigenvalues are distributed according to a density function that is a perfect semicircle! The size is $R \sim \sqrt{Ng}$ (technically speaking, Wigner did this for Hermitian matrices).

What does this have to do with holography? Let $\rho(\lambda)$ denote the density of eigenvalues. It turns out that we obtain in the large N limit approximately

$$\alpha\rho^2 + \lambda^2 = gN.$$

Therefore we have the density function

$$\rho(\lambda) \propto \sqrt{R^2 - \lambda^2}.$$

But now there is the new variable ρ, which represents a new dimension. See Fig.91. In this sense, holography has to do with understanding some property at the extremes, in this case when the size of the matrix N is large. This gives an equivalent picture: the semicircle law. In more complicated examples in string theory, it gives gravity in one higher dimension. These pictures can be very illuminating and interesting in their own right.

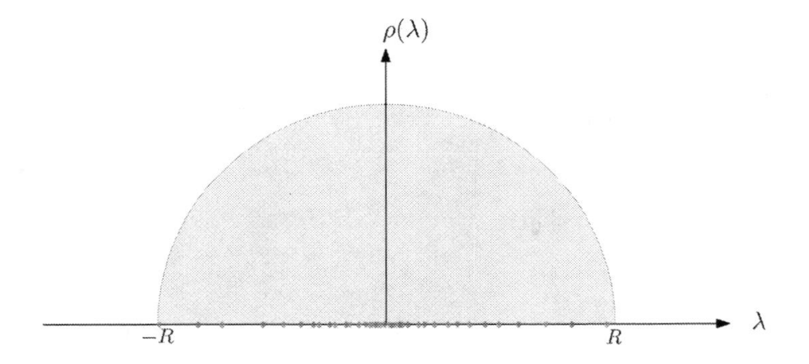

FIGURE 91. The density of eigenvalues of a large random Hermitian matrix will appear as a semi-circle.

Holography, in the sense we are talking about here, is an amazing idea discovered through string theory–an incredible puzzle solver that we can bring to bear on a variety of problems. We do not always know why this and other approaches based on dualities work. Understanding that will require a long-term collaboration between mathematicians and physicists. No one yet knows how long a project like that will take nor where it will ultimately lead us.

12. **Summing Up**

Many of our discussions in previous chapters have alternated between physics and mathematics, though in the last chapter we saw them intertwined in the concept of *duality*. In some cases, we have discussed a given topic (A) from the mathematical as well as physical perspective. We then discussed the opposite topic (let's call it anti-A), again from both perspectives. We also worked through some puzzles that combined ideas from physics and math into a kind of dual picture.

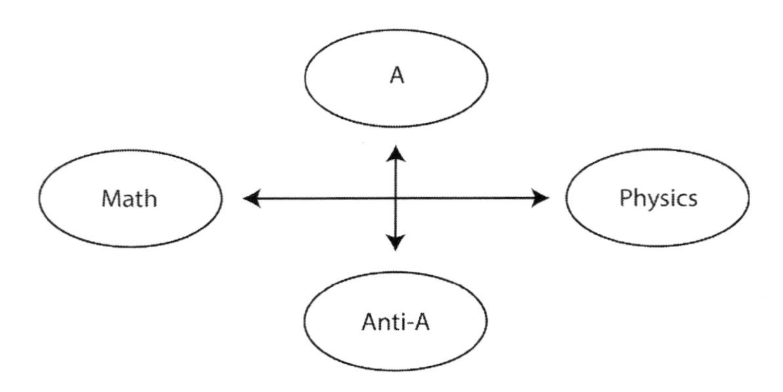

FIGURE 92. A rough diagram of the topics of the lectures.

Duality, as we have explained, has become a theme of growing importance in physics today, and it is also an important idea for summarizing what we have learned in these lectures. One lesson gleaned from an exploration of the dualities that emerge in nature is that we must be open to diverse ideas in our consideration of physical laws. We should not hold fast to a single perspective, disregarding alternative viewpoints. There are many different ways of looking at things: all may be equally legitimate and all may offer their own advantages and insights. In some cases, a better perspective can give rise to a better solution. This, in a nutshell, may be the key, overriding message to be drawn from our discussions. Let us review some of the things we have covered, while adding a few details here and there.

Symmetries and Their Breaking

We first looked into the importance of symmetries in physics. We looked at translational symmetries and rotational symmetries, as well as other, more subtle forms of symmetry.

Puzzle

What is the measure of the indicated angle where the two indicated sides are correspondingly equal?

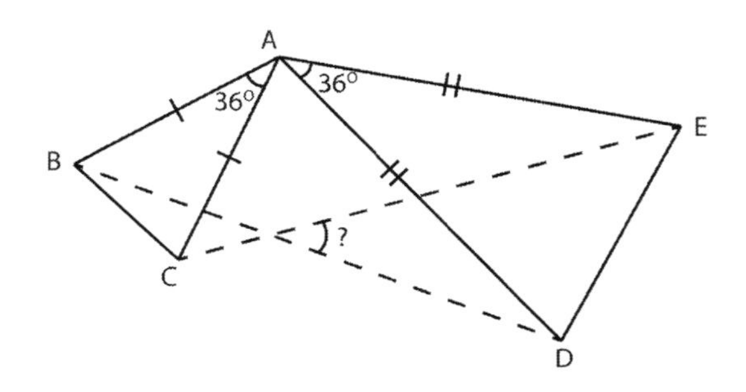

Solution

We see that $\triangle ABD \cong \triangle ACE$ are equal triangles and, after a rotation of 36 degrees about A, they become identical. So, therefore, the angle between BD and CE is 36° because that is how much the angles between the lines change during this rotation. Our solution also makes use of the idea of rotational symmetry, because we know the triangle can be rotated (by 36° in this case) without changing in any other way.

Puzzle

There are four turtles at the corners of a square of size 10 m. Each moves towards the adjacent turtle counterclockwise at a constant speed of 1 m/s, and takes the shortest path towards its target. How long does it take them to meet at the center?

Solution

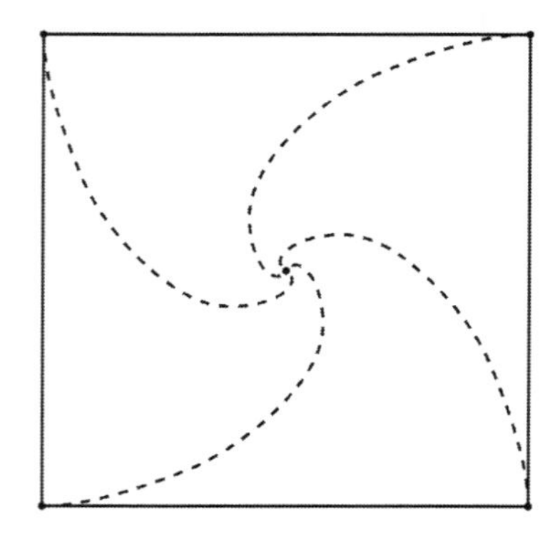

It takes 10 seconds. Intuition tells us that from one turtle's perspective, if the other turtle were not moving, it would takes 10 seconds. But why is it true after the target turtle is also moving?

The system always possesses 90° rotational symmetry about the origin, so the turtles *always* form a configuration of a square as they all move (see Fig.93). The turtles start out at the four corners of a square, and always maintain the same distance between each other, which means they're always on the four corners of a square. Therefore, the chasing turtle is *always* moving perpendicular to the chased turtle's direction of motion, so the rate of approach is the same as if the chased turtle is not moving.

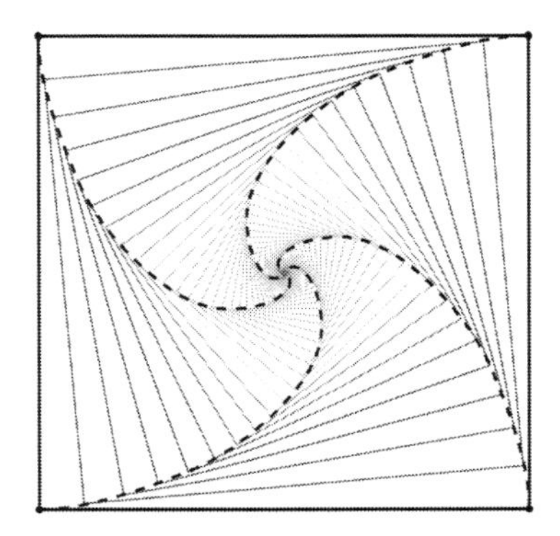

FIGURE 93. As the turtles move they are always at the corners of a square.

Symmetry is fun but breaking it is sometimes even more fun.

In some ways, we owe our existence to symmetry breaking. Matter is dual to anti-matter, so how does matter exist whereas anti-matter has disappeared? If it were a perfectly symmetric relation, then no matter would exist, and we would not be here. Matter and anti-matter would annihilate one another, leaving nothing behind. Having leftover matter is only possible because of a tiny discrepancy between matter and anti-matter (due to only one matter particle out of roughly 10^9 imbalance), called the "CP violation." Even though we say that symmetry is all around us, symmetry breaking is around us too, and it may be even more important.

Symmetry breaking, however, can be very unintuitive. For instance, physics is not symmetric via reflection, a fact that even Feynman found difficult to accept at first.

Gauge Symmetry

Many important properties of particle physics involve what are called "gauge symmetries." These have a somewhat different flavor from the more familiar symmetries we see around us. With regard to translational symmetry, we might say that an experiment performed at two different points should have the same result. With regard to gauge symmetry, we might say that these two different points are essentially the *same* point. Mathematically, we would say that we are "modding out" by some equivalence relation. One example of modding out would be to show the equivalence between all horizontal lines on the cylinder. A cylinder can be thought of as the product of a line and a circle. If every point of a circle labels a line (namely the one that passes through it). Consider a gauge symmetry which is rotating the cylinder along its circumference. If we identify this as a "gauge symmetry" we would be identifying all the points of the circle with each other. In other words, all the horizontal lines of the cylinder would be viewed as equivalent.

Classical electromagnetism provided the first known example of a gauge symmetry which realizes the gauge symmetry of the cylinder discussed above. For every electric field, there is an electric potential (V) assigned to every point in space. It turns out that the numerical value of V at any given point is arbitrary because it is defined relative to a reference point or "ground" that is, itself, arbitrary. If one were to choose a different reference point or ground, the numerical value of V would shift, but that would only change the units without affecting the physics of the situation. Put in other terms, if V is a solution to Maxwell's equations, then V plus some arbitrary constant, C, would also solve the equations, and there would be no discernible change to the electric and magnetic fields. This is the simplest example of a gauge transformation. Maxwell's equations, in physics parlance, have a gauge symmetry because the equations are invariant under a more complicated version of such transformation, which is referred to as a gauge transformation. In the analogy with the cylinder, it is equivalent to choosing a point on the circle for each point in space-time.

The practice of currency exchange between countries offers a good analogy of gauge symmetry. We can imagine a lattice of dots, with each dot representing a country. And between each neighboring country there is an exchange rate for currency.[40]

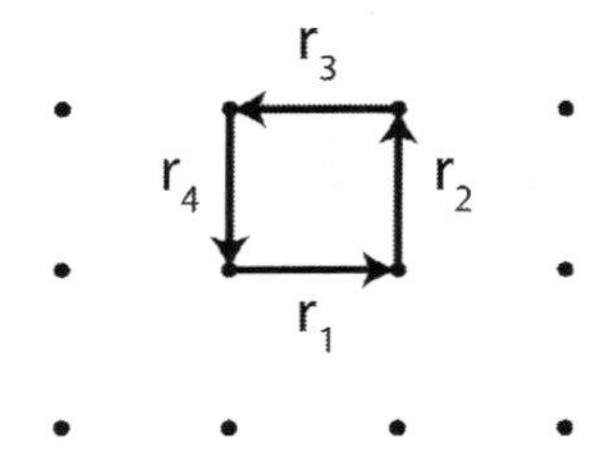

Changing the unit of currency in one country will change all the exchange rates r_i between that country and other ones. Such a change in the unit of currency does not actually affect any economic factor, as it is simply a matter of convention. Consider an example of currency exchange between the United States and Argentina. Suppose that the exchange rate had been $3,000 (Argentine) pesos = 1 (U.S.) dollar$. In 1985, Argentina introduced the Austral as its standard unit of currency. If $1 Austral = 1,000 pesos$, then a U.S. dollar would then be worth 3 Australes. This is called a gauge symmetry because

[40]For a more detailed discussion of this example see the paper by Juan Maldacena: https://arxiv.org/pdf/1410.6753.pdf .

after this conversion change, nothing has really changed. All currency unit choices are equivalent. In that sense it is like a gauge symmetry. Note that the product $r_1 r_2 r_3 r_4$ around a loop will *not* change due to shifts in the unit of currency, and it is physically meaningful. It is what we physicists call a gauge invariant quantity. There can either be a fair exchange rate, where cycles multiply out to unity ($r_1 r_2 r_3 r_4 = 1$). Or, in other cases, speculation can allow people to actually make money through arbitrage ($r_1 r_2 r_3 r_4 > 1$).

This is a simple picture of gauge symmetries, though it does not give any indication of their crucial role in physics. Electromagnetism, as discussed earlier, has a $U(1)$ gauge symmetry. Our theories of the three other fundamental forces of nature–weak, strong, and gravitational–also incorporate gauge symmetries, so it is a central construct upon which our Standard Model of particle physics firmly rests.

Intuitive Math

We have discussed a variety of mathematical phenomena in physics. For instance, we were able to argue that there are antipodal points on Earth with the same temperature and air pressure, based upon nothing but the principle of continuity. We showed through this example that math can put some constraints on physics, though it often does not give much information about the guts and dynamics of physics. Math can lay down the constraint regarding the existence of those antipodal points on Earth, but it does not tell you where those points are or the circumstances under which that might change.

Puzzle

Imagine that we have a ball whose surface is filled with atoms. We make each atom move around on the surface for a while, continuously, and then settle down (like in musical chairs). An atom "loses" if it ends up in the same spot. The atoms want to help each other not to lose. Is it possible for no atom to lose?

Solution

No, this is not possible! A mathematical theorem, called Brouwer's fixed point theorem, holds that a continuous map of a sphere into itself always has a fixed point. The proof is similar to the arguments concerning winding numbers that we discussed earlier. The infinitesimal version of this gives a vector field, which must have a zero somewhere. This phenomenon is also described by the assertion: You can't comb the hair at every point on a sphere. There is always a spot where the hair sticks straight up and cannot be combed in any direction you choose. Projecting the hairs onto the sphere gives a vector and the one sticking up gives a zero vector and that is what we mean by a fixed point of that vector field flow. Movement of the atoms in the above puzzle, can be viewed as a vector field, and for no atom to lose the vector field should not be zero anywhere. This is impossible.

Again, these seemingly formal mathematical statements play important roles in physics. The fact that the winding number is conserved upon concatenating curves around a cylinder corresponds to conservation of charge, as we have discussed before.

There is a Skyrme model for protons, which goes as follows. For starters, there is a field g that takes values on S^3. The value of this field over the space $\mathbb{R}^3$ can be viewed as a map $g : \mathbb{R}^3 \to S^3$, where R^3 is our space and we call g a field. A proton (more generally, a baryon) can be viewed under the following framework. We consider a one-point compactification of the space, replacing R^3 by S^3 by adding a point at infinity. A single baryon corresponds to the identity map $g : S^3 \to S^3$. More generally, the winding number of this map gives the baryon number.

Counterintuitive Math

A prime example of counterintuitive math is the approximate 120 m of extra height that can be realized by adding 1 m of rope around the equator of the earth.

Recall the puzzle we had about the number of regions in a disk cut out by lines between points on the boundary. We learned that we can be deceived by patterns. In physics, we will only ever have finite data and examples from which to extrapolate.

Even if our theories fit our finite data perfectly, we must always be prepared for the possibility that they are wrong.

Infinities are often a source of confusion and puzzlement. An interesting aspect of this is that the set of computable real numbers–those we can calculate to as many digits as we want–is countable, despite the fact that real numbers, as a whole, are uncountable. This is because computable real numbers are made up of a countable set of operations. And if you do a countable number of operations, unsurprisingly, you will get a countable number of results.

Sometimes, crazy math actually shows up in physics. Recall our previous discussion of Hilbert's Hotel, $\infty + 1 = \infty$. This actually happens in physics to produce a particle in a vacuum. The mathematical theory, which describes these kinds of anomalies, is called *index theory.*

Puzzle

We are flipping a special coin. The first flip comes up heads, and the second flip comes up tails. The coin has the following adaptive property: After the first two flips, the probability of flipping heads is proportional to the ratio of heads that have turned up in the previous flips.

We do 100 more flips. What is the probability of flipping 13 heads (after the first two fixed flips)? 50 heads? 100 heads?

Solution

There are two competing effects. There is the combinatoric aspect regarding the number of ways reaching a specific number of heads and an avalanche effect, which favors the same flipping result as the previous ones. The first effect would make 50 heads much more likely than 0 or 100 heads. The second effect would do the opposite and make the extremes very likely. Say per chance you got heads for the first few flips, then the chances to get heads increases and it could lead to an avalanche effect of heads, making the more extreme choices likely. It turns out that the two effects actually cancel, and the result is that the probability does not depend on the number of heads! All the outcomes are equally likely with probability $p = \frac{1}{101}$.

We can see this by looking at the following sequence of flips and the corresponding probability of each flip.

H	T	H	H	T	H	T
		$\frac{1}{2}$	$\frac{2}{3}$	$\frac{1}{4}$	$\frac{3}{5}$	$\frac{2}{6}$

From this pattern we can deduce that for a given result of H heads and T tails, the probability is

$$\binom{H+T}{H} \times \frac{H!T!}{(H+T+1)!} = \frac{1}{H+T+1}$$

The first factor relates to the combinatoric effect and the second factor to the avalanche effect. This leads to probability $1/101$ for all outcomes.

Intuitive and Unintuitive Physics

Physical intuition can be used to gain surprising insight into seemingly difficult problems. A prominent example of this is that physical intuition, by way of string theory, has led to progress in abstract problems (of enumerative geometry) that were of wholly independent mathematical interest. At the frontiers of theoretical physics, we have used mirror symmetry and physical intuition to predict mathematical phenomena that mathematicians, themselves, cannot yet prove.

Recall how we proved the Pythagorean theorem using an argument based on torques? One can argue about whether or not this argument was somehow circular, but it is unquestionable that casting mathematical problems in a physical framework can offer new insights. For instance, you can readily use this technique to see the more general law of cosines. These are just a couple of easy, toy examples that help make the more general point.

Theories often do not appear intuitive until you view them through the right lens. For instance, Einstein's theory of relativity seems very unintuitive, with its strange phenomena of time dilation and length contraction, but it stems from the very intuitive assumption that inertial frames must be identical. At first, nobody thought that the speed of light would be the same in all inertial frames, and physicists spent many years searching for an ether–the medium through which they thought light propagated–so they could measure the speed of light in different frames. But these fruitless efforts merely led to the conclusion that the postulated ether does not exist. And once you accept that the speed of light is the same for all reference frames, Einstein's theory immediately follows, and it becomes more "intuitive."

Not all physics is intuitive on the other hand. A prime example of unintuitive physics is quantum mechanics. Alas, we have not yet found the right light to view quantum mechanics. It is a very unintuitive theory. Quantum mechanics, and especially the probabilistic aspects of it, is still elusive to our minds. One hundred years have passed, and we still have not internalized it. An example of this is the double slit experiment, which still seems contrary to common sense, after so many years. Other unintuitive aspects of quantum mechanics include the inability to tag identical particles such as electrons–an issue we grappled with in an earlier puzzle about ants.

Naturalness

We then moved on to naturalness where we saw that we could basically compute many things using dimensional analysis. Our answers could be off by moderate amounts through this approach, but they were still generally $O(1)$.

This, at least, is what physicists used to think. Dirac was one of the first to point out that there are some huge numbers occurring naturally in physics. For instance, the ratio of the electromagnetic repulsion to the gravitational attraction between two protons is a dimensionless quantity, which turns out to be astronomically large. Using $\hbar, G$, and c, we found fundamental units of nature, so-called Planck units.

From the Planck length to the length of the universe we have

$$\ell_{\text{Planck}} \xrightarrow{\times 10^{20}} \ell_{\text{proton}} \xrightarrow{\times 10^{20}} \ell_{\text{Sun}} \xrightarrow{\times 10^{20}} \ell_{\text{Universe}}$$

We encountered these hierarchy puzzles, where unnaturally big and small numbers emerge. Why this hierarchy of scales exists is an open question. A more natural arrangement would probably have led to a universe on the order of the Planck scale! If that's the case, our very existence would have to be regarded as highly unlikely, as well as unnatural. Under more natural conditions, the universe would only exist on a Planck time scale, or just a very tiny fraction of a second. Some physicists have invoked the anthropic principle in an attempt to make fine tuning appear natural, but that approach has led to very few predictions–Weinberg's successful prediction of the cosmological constant Λ (based on the anthropic principle) being one of the rare bright spots.

We discussed how natural problems in number theory, formulated in terms of small numbers, can nevertheless lead to huge numbers as their solutions. Perhaps the hierarchy issue that seems unnatural in a physical context has a similar explanation: It may just be a matter of posing the question in the right way in terms of variables that are $O(1)$.

Physics and Religion

We discussed some philosophical ideas related to physics and religion. One source of mystery is that the parameters of the universe need to be very finely tuned in order for us to even exist. Creationists would say that someone, or some being, had to tune these parameters–by hand, as it were. People who believe in more natural explanations would appeal to the anthropic principle, suggesting that our universe has been evolutionarily selected for us to exist, or else we wouldn't even be here to ponder such questions. We ultimately concluded that science cannot disprove religion, nor can religion disprove science. We should not draw conclusions from one school of thought and apply them to the other. When you go back in time–say, to Newton's era in the 1600s–you find varying shades of religiosity. I think scientists are religious, in perhaps unconventional senses, even though they may not acknowledge it. They are looking for a pattern that may not even exist, but they believe that it does, which might be considered irrational and somewhat akin to a religious belief. The important thing is to keep an open mind. Without it, Einstein remained skeptical about the existence of black holes and gravitational waves–even though they were predicted by his theory of general relativity. And because of his preconceived notions, Einstein dismissed the Big Bang theory as Christian mythology. Despite those stances Einstein's position makes him one of the greatest visionaries in the history of science.

Duality

We saw how dualities–which seem to naturally arise in physics and mathematics–can transform a hard question into an easy question. This is a revolutionary concept, whose impact on physics is already huge, and which continues to extend its reach ever further into mathematics. However, it is almost embarrassing that we don't understand why these ideas work. There may, however, be a philosophical reason for why dualities should exist in nature: Physical theories are so rich with structures that it almost seems as if you would have to postulate too many miraculous things for them to exist. So if you get the same miracles working for two different looking theories, perhaps they are really the same in disguise, i.e. they are dual to one another. This is Sergio Cecotti's explanation of why we have dualities: The scarcity of rich structures forces many of them to be repeats!

Advances in theory are changing our conception of the universe and challenging our grasp of fundamental notions like mass, and the space-time we supposedly inhabit. That said, there may be limits as to how far theory can take us. Over the past several decades, our theoretical frontiers have pushed far ahead of our experimental capabilities, and we have not been able to translate many of our latest discoveries into observational data. Nevertheless, ideas from theoretical physics continue to find applications in mathematics, which has turned out to be a very productive avenue. Connections between these fields are growing stronger as science pushes forward in new, unexpected ways. Even though we don't know the exact itinerary, it is certain to be a thrilling ride. I hope you join us on this adventure!

Index

Made in the USA
Middletown, DE
02 September 2024